Nondestructive Testing in Composite Materials

Nondestructive Testing in Composite Materials

Editor

Carosena Meola

MDPI • Basel • Beijing • Wuhan • Barcelona • Belgrade • Manchester • Tokyo • Cluj • Tianjin

Editor
Carosena Meola
Università di Napoli Federico II
Italy

Editorial Office
MDPI
St. Alban-Anlage 66
4052 Basel, Switzerland

This is a reprint of articles from the Special Issue published online in the open access journal *Applied Sciences* (ISSN 2076-3417) (available at: https://www.mdpi.com/journal/applsci/special_issues/Nondestructive_Testing_in_Composite_Materials).

For citation purposes, cite each article independently as indicated on the article page online and as indicated below:

LastName, A.A.; LastName, B.B.; LastName, C.C. Article Title. *Journal Name* **Year**, *Volume Number*, Page Range.

ISBN 978-3-03943-731-3 (Hbk)
ISBN 978-3-03943-732-0 (PDF)

Cover image courtesy of Carosena Meola.

© 2020 by the authors. Articles in this book are Open Access and distributed under the Creative Commons Attribution (CC BY) license, which allows users to download, copy and build upon published articles, as long as the author and publisher are properly credited, which ensures maximum dissemination and a wider impact of our publications.

The book as a whole is distributed by MDPI under the terms and conditions of the Creative Commons license CC BY-NC-ND.

Contents

About the Editor vii

Carosena Meola
Nondestructive Testing in Composite Materials
Reprinted from: *Appl. Sci.* **2020**, *10*, 5123, doi:10.3390/app10155123 1

Milad Mosharafi, SeyedBijan Mahbaz and Maurice B. Dusseault
Simulation of Real Defect Geometry and Its Detection Using Passive Magnetic Inspection (PMI) Method
Reprinted from: *Appl. Sci.* **2018**, *8*, 1147, doi:10.3390/app8071147 5

Junsheng Zhang, Zhijie Guo, Tengyun Jiao and Mingquan Wang
Defect Detection of Aluminum Alloy Wheels in Radiography Images Using Adaptive Threshold and Morphological Reconstruction
Reprinted from: *Appl. Sci.* **2018**, *8*, 2365, doi:10.3390/app8122365 25

Nobuyuki Toyama, Jiaxing Ye, Wataru Kokuyama and Shigeki Yashiro
Non-Contact Ultrasonic Inspection of Impact Damage in Composite Laminates by Visualization of Lamb wave Propagation
Reprinted from: *Appl. Sci.* **2019**, *9*, 46, doi:10.3390/app9010046 37

Alessandro Grazzini
In Situ Analysis of Plaster Detachment by Impact Tests
Reprinted from: *Appl. Sci.* **2019**, *9*, 258, doi:10.3390/app9020258 47

Guoyang Teng, Xiaojun Zhou, Chenlong Yang and Xiang Zeng
A Nonlinear Method for Characterizing Discrete Defects in Thick Multilayer Composites
Reprinted from: *Appl. Sci.* **2019**, *9*, 1183, doi:10.3390/app9061183 59

Hossein Taheri and Ahmed Arabi Hassen
Nondestructive Ultrasonic Inspection of Composite Materials: A Comparative Advantage of Phased Array Ultrasonic
Reprinted from: *Appl. Sci.* **2019**, *9*, 1628, doi:10.3390/app9081628 75

Qi Zhu, Yuxuan Ding, Dawei Tu, Haiyan Zhang and Yue Peng
Experimental Study of Defect Localization in a Cross-Ply Fiber Reinforced Composite with Diffuse Ultrasonic Waves
Reprinted from: *Appl. Sci.* **2019**, *9*, 2334, doi:10.3390/app9112334 91

Carlo Boursier Niutta, Andrea Tridello, Raffaele Ciardiello, Giovanni Belingardi and Davide Salvatore Paolino
Assessment of Residual Elastic Properties of a Damaged Composite Plate with Combined Damage Index and Finite Element Methods
Reprinted from: *Appl. Sci.* **2019**, *9*, 2579, doi:10.3390/app9122579 103

Ping Zhou, Gongbo Zhou, Zhencai Zhu, Zhenzhi He, Xin Ding and Chaoquan Tang
A Review of Non-Destructive Damage Detection Methods for Steel Wire Ropes
Reprinted from: *Appl. Sci.* **2019**, *9*, 2771, doi:10.3390/app9132771 117

Simone Boccardi, Natalino Daniele Boffa, Giovanni Maria Carlomagno, Giuseppe Del Core, Carosena Meola, Ernesto Monaco, Pietro Russo and Giorgio Simeoli
Lock-In Thermography and Ultrasonic Testing of Impacted Basalt Fibers Reinforced Thermoplastic Matrix Composites
Reprinted from: *Appl. Sci.* **2019**, *9*, 3025, doi:10.3390/app9153025 . . . . **133**

Wongi S. Na and Ki-Tae Park
Toward Creating a Portable Impedance-Based Nondestructive Testing Method for Debonding Damage Detection of Composite Structures
Reprinted from: *Appl. Sci.* **2019**, *9*, 3189, doi:10.3390/app9153189 . . . . **145**

Hanchao Li, Yating Yu, Linfeng Li and Bowen Liu
A Weighted Estimation Algorithm for Enhancing Pulsed Eddy Current Infrared Image in Ecpt Non-Destructive Testing
Reprinted from: *Appl. Sci.* **2019**, *9*, 4199, doi:10.3390/app9204199 . . . . **155**

About the Editor

Carosena Meola, aeronautical engineer, is a senior research staff member at the Department of Industrial Engineering/Aerospace Division—University of Naples Federico II. Meola has attained Level III in infrared thermography and is a licensed instructor for personnel training and certification. Meola is a member of UNI, CEN and ISO Technical Committees in addition to serving on the editorial board of numerous international journals and on the scientific committee of international conferences as well as Chair of conference sessions, Editor of numerous books, and Guest Editor of journal Special Issues. Meola is author and co-author of around 200 papers published in well recognized journals, books, and proceedings and serves as referee for around 50 international journals and projects.

Editorial

Nondestructive Testing in Composite Materials

Carosena Meola

Department of Industrial Engineering, University of Naples Federico II, 80125 Napoli, Italy; carmeola@unina.it

Received: 16 June 2020; Accepted: 18 July 2020; Published: 25 July 2020

1. Introduction

A composite material is made of two or more constituents of different characteristics with the intent to complete the shortcomings of the individual components and to get a final product of specific characteristics and shape [1] to fulfil the user's demand. The most extraordinary example of composite is found in nature; in fact wood, which appears so strong and resistant, is composed of long fibers of cellulose held together by the lignin that is a weaker substance. Human beings observing and copying nature have always strived to develop composite materials. An example of composite material comes from afar: mud bricks; these were created when the ancients realized that mixing mud and straw gave them a resistant building material such as mud bricks. Later on, concrete was originated from the combination of cement, sand and gravel, and was widely used in the construction sector. Many types of materials have been developed and continue to be developed to meet the different needs of the modern world. Different types of matrices and reinforcements are being used that are derived from petrochemical resources or extracted from the vegetable world [2], which also allows us to comply with safety at work concerns and waste disposal. Indeed, the combination of two elements represents for many composite materials a strength and weakness at the same time. In fact, several different types of defects [3] may occur during the fabrication of composites, with the most common being: fiber/play misalignment, broken fibers, resin cracks or transversal ply cracks, voids, porosity, slag inclusions, non-uniform fiber/resin volume ratio, disbonded interlaminar regions, kissing bonds, incorrect cure and mechanical damage around machined holes and/or cuts. The presence of defects may result in a considerable drop of the composite mechanical properties [4]. Therefore, effective non-destructive evaluation methods able to discover defects at an incipient stage are necessary to either assure the quality of a composite material prior to putting it into service, or to monitor a composite structure in service.

2. Nondestructive Testing

We all would like to live in a safe house that would not collapse on us. We would all like to walk on a safe road and never see a chasm open in front of us. We would all like to cross a bridge and reach the other extreme safely. We all would like to feel safe and secure to take the plane, the ship, the train or to use any equipment. All this may be possible with the adoption of adequate manufacturing processes, non-destructive inspection of final parts and monitoring during the in-service life. This requires effective non-destructive testing techniques and procedures. The intention of this special issue was to collect the latest research to highlight new ideas and the way to deal with challenging issues worldwide. There were 19 papers submitted of which 12 were accepted and published. Going through the special issue, different types of materials and structures were considered; different non-destructive testing techniques were employed with new approaches of data treatment proposed as well numerical simulation.

The degradation of concrete, the material of which many widely used goods are made of such as roads, bridges and the home in which we live, is certainly a cause of anxiety and demands for safety. Milad Mosharafi, SeyedBijan Mahbaz and Maurice B. Dusseault dealt with the problem of corrosion of steel in reinforced concrete [5]. The authors reviewed previous literature and focused on the self-magnetic behavior of ferromagnetic materials, which can be exploited for quantitative condition assessment. In particular, they performed numerical simulation to get information on the possibility to detect the rebar degradation with the passive magnetic inspection method and to establish detectability limits of such method. Of great relevance for all us is the safeguard of the cultural heritage, which represents our history; the paper by Grazzini [6] can be inserted in this context. Grazzini describes a technique to detect plaster detachments from historical wall surfaces that consist of small and punctual impacts exerted with a specific hammer on the plastered surface. This technique was applied to frescoed walls of Palazzo Birago in Turin (Italy).

Most of the papers of this special issue involve fiber reinforced composites [7–12]. These include different types of matrices and fibers that are used for different applications going from the transport industry (aircraft, trains, ships, etc.) to goods for daily life. The most popular are those based on resin epoxy matrix reinforced with either carbon or glass fibers and are named CFRP for carbon fiber reinforced polymer and GFRP for glass fiber reinforced polymer; these materials are also called carbon/epoxy and glass/epoxy. These materials can be non-destructively evaluated by using different techniques, amongst them ultrasonic testing (UT) and infrared thermography (IRT). Ultrasonic testing in reflection mode (pulse-echo) can be accomplished with a single probe (SEUT), which acts to both send and receive sound waves, or with a phased array (PAUT). The superiority in terms of the signal noise ratio of PAUT over SEUT was assessed by Hossein Taheri and Ahmed Arabi Hassen through a comparative study on a GFRP sample [7]. The authors of Ref. [7] used the same PAUT for guided wave generation to detect flaws in a CFRP panel.

In addition to the use of the direct wave, the diffuse wave can also be exploited for inspection purposes. The information contained in diffuse waves are mostly useful in seismology and in civil engineering, but can be also used for health monitoring and the nondestructive evaluation of fiber reinforced composites. Zhu et al. [8] applied this method to the inspection of carbon/epoxy and found it promising for early crack detection. A critical aspect for defect localization is to distinguish signals from noise, and this requires more investigation.

Carbon fiber reinforced polymer laminates are also considered by Toyama et al. [9]. The latter authors used non-contact ultrasonic inspection technique through visualization of Lamb wave propagation for detecting barely visible impact damage in CFRP laminates. Ultrasonic testing is generally a contact technique, but this poses problems in materials and structures in which the contact fluid (water, gel) may be hurtful for the surface; thus, the non-contact deployment is of great interest and ever more investigated. The results reported in Ref. [9] are promising but, as also concluded by the authors, the method based on Lamb waves requires further investigation with particular regard to the signal-to-noise ratio improvement.

Teng et al. [10] investigated the suitability of the recurrence quantification analysis in ultrasonic testing to characterize small size defects in a thick, multilayer, carbon fiber reinforced polymer. The authors conclude that their proposed method was able to detect artificial defects in the form of blind holes, but further research is necessary to improve and update the method to address real discrete defects. Niutta et al. [11] used the detecting damage index technique in combination with the finite element method to evaluate residual elastic properties of carbon/epoxy laminates damaged through repeated four-point bending tests. As a conclusion, the authors of Ref. [11] affirm that their methodology allows us to locally assess the residual elastic properties of damaged composite materials. By mapping the elastic properties on the component and considering the assessed values in a finite element model, a precise description of the mechanical behavior of the composite plate is obtained and, consequently, the health state of a damaged component can be quantitatively evaluated and decisions on its maintenance can be made by defining limits on the acceptable damage level.

Infrared thermography is widely used in the inspection of materials and structures, amongst them composites, thanks to its remote deployment through the use of a non-contact imaging device. Lock-in thermography coupled with ultrasonic phased array was used by Boccardi et al. [12] to detect impact damage in basalt-based composites. In particular, two types of materials that include basalt fibers as reinforcement of two matrices were considered: polyamide and polypropylene. The obtained results show that both techniques can discover either impact damage or manufacturing defects. However, lock-in thermography, being non-contact, can be used with whatever surface while contact ultrasonic cannot be used on hydrophilic surfaces that get soaked with the coupling gel. Infrared thermography lends itself to being integrated with other techniques to allow the inspection of both thin and thick structures such as in Ref. [13], in which a joint use of infrared thermography with a ground penetrating radar (GPR) allowed us to assess the conditions of archaeological structures.

In particular, IRT was able to detect shallow anomalies while the GPR followed their evolution in depth. The integration of infrared thermography with other techniques is also deployed with IRT for the detection of defects, and the other technique is exploited for thermal stimulation. An example of this deployment is ultrasound thermography [14], in which elastic waves are used for selective heating and infrared thermography detects buried cracks. An example of integration between infrared thermography and eddy current is given by Li et al. in Ref. [15] of this special issue, in which the pulsed eddy current is used for thermal stimulation to detect welding defects.

The paper by Zhang et al. [16] is concerned with a technical solution that combines the adaptive threshold segmentation algorithm and the morphological reconstruction operation to extract the defects on wheel X-ray images. The obtained results show that this method is capable of accurate segmentation of wheel hub defects. The authors claim that the method may be suitable for use in other applications, but warn about the importance of using the proper parameter settings. Na and Park [17] investigated the possibility to transform the electromechanical impedance (EMI) technique into a portable system with the piezoelectric (PZT) transducer temporarily attached and detached by using a double-sided tape. Regardless of the damping effect, which may cause the impedance signatures to be less sensitive when subjected to damage, the results from this study have demonstrated its feasibility. The authors are convinced that, by conducting simulation studies, the PZT size can be further reduced for a successful debonding detection of composite structures.

At last, Zhou et al. [18] made an overview of nondestructive methods for the inspection of steel wire ropes. The authors first analyzed the causes of damage and breakage as local flaws and the loss of the metallic cross-sectional area. Then, they reviewed several detection methods, including electromagnetic detection, optical detection, ultrasonic guided wave method, acoustic emission detection, eddy current detection and ray detection, by considering the advantages and disadvantages. They found that the electromagnetic detection method has gradually been applied in practice, and the optical method has shown great potential for application, while other methods are still in the laboratory stage.

Funding: This research received no external funding.

Conflicts of Interest: The authors declare no conflict of interest.

References

1. Hull, D.; Clyne, T.W. *An Introduction to Composite Materials*, 3rd ed.; Cambridge University Press: Cambridge, UK, 2019.
2. Mohanty, A.K.; Misra, M.; Drzal, L.T. Sustainable Bio-Composites from Renewable Resources: Opportunities and Challenges in the Green Materials World. *J. Polym. Environ.* **2002**, *10*, 19–26. [CrossRef]
3. Smith, R.A. Composite Defects and Their Detection, Materials Science and Engineering, Vol. III. Available online: https://www.eolss.net/Sample-Chapters/C05/E6-36-04-03.pdf (accessed on 2 March 2020).
4. Summerscales, J. Manufacturing Defects in Fibre Reinforced Plastics Composites. *Insight* **1994**, *36*, 936–942.
5. Mosharafi, M.; Mahbaz, S.B.; Dusseault, M.B. Simulation of Real Defect Geometry and Its Detection Using Passive Magnetic Inspection (PMI) Method. *Appl. Sci.* **2018**, *8*, 1147. [CrossRef]
6. Grazzini, A. In Situ Analysis of Plaster Detachment by Impact Tests. *Appl. Sci.* **2019**, *9*, 258. [CrossRef]
7. Taheri, H.; Hassen, A.A. Nondestructive Ultrasonic Inspection of Composite Materials: A Comparative Advantage of Phased Array Ultrasonic. *Appl. Sci.* **2019**, *9*, 1628. [CrossRef]

8. Zhu, Q.; Ding, Y.; Tu, D.; Zhang, H.; Peng, Y. Experimental Study of Defect Localization in a Cross-Ply Fiber Reinforced Composite with Diffuse Ultrasonic Waves. *Appl. Sci.* **2019**, *9*, 2334. [CrossRef]
9. Toyama, N.; Ye, J.; Kokuyama, W.; Yashiro, S. Non-Contact Ultrasonic Inspection of Impact Damage in Composite Laminates by Visualization of Lamb wave Propagation. *Appl. Sci.* **2019**, *9*, 46. [CrossRef]
10. Teng, G.; Zhou, X.; Yang, C.; Zeng, X. A Nonlinear Method for Characterizing Discrete Defects in Thick Multilayer Composites. *Appl. Sci.* **2019**, *9*, 1183. [CrossRef]
11. Niutta, C.B.; Tridello, A.; Ciardiello, R.; Belingardi, G.; Paolino, D.S. Assessment of Residual Elastic Properties of a Damaged Composite Plate with Combined Damage Index and Finite Element Methods. *Appl. Sci.* **2019**, *9*, 2579. [CrossRef]
12. Boccardi, S.; Boffa, N.D.; Carlomagno, G.M.; Del Core, G.; Meola, C.; Monaco, E.; Russo, P.; Simeoli, G. Lock-In Thermography and Ultrasonic Testing of Impacted Basalt Fibers Reinforced Thermoplastic Matrix Composites. *Appl. Sci.* **2019**, *9*, 3025. [CrossRef]
13. Carlomagno, G.M.; Di Maio, R.; Fedi, M.; Meola, C. Integration of infrared thermography and high-frequency electromagnetic methods in archaeological surveys. *J. Geophys. Eng.* **2011**, *8*, S93–S105. [CrossRef]
14. Zweschper, T.; Dillenz, A.; Riegert, G.; Busse, G. Ultrasound Thermography in NDE: Principle and Applications. In *Acoustical Imaging*; Arnold, W., Hirsekorn, S., Eds.; Springer: Dordrecht, The Netherlands, 2004; Volume 27, pp. 113–120.
15. Li, H.; Yu, Y.; Li, L.; Liu, B. A Weighted Estimation Algorithm for Enhancing Pulsed Eddy Current Infrared Image in Ecpt Non-Destructive Testing. *Appl. Sci.* **2019**, *9*, 4199. [CrossRef]
16. Zhang, J.; Guo, Z.; Jiao, T.; Wang, M. Defect Detection of Aluminum Alloy Wheels in Radiography Images Using Adaptive Threshold and Morphological Reconstruction. *Appl. Sci.* **2018**, *8*, 2365. [CrossRef]
17. Na, W.S.; Park, K.-T. Toward Creating a Portable Impedance-Based Nondestructive Testing Method for Debonding Damage Detection of Composite Structures. *Appl. Sci.* **2019**, *9*, 3189. [CrossRef]
18. Zhou, P.; Zhou, G.; Zhu, Z.; He, Z.; Ding, X.; Tang, C. A Review of Non-Destructive Damage Detection Methods for Steel Wire Ropes. *Appl. Sci.* **2019**, *9*, 2771. [CrossRef]

© 2020 by the author. Licensee MDPI, Basel, Switzerland. This article is an open access article distributed under the terms and conditions of the Creative Commons Attribution (CC BY) license (http://creativecommons.org/licenses/by/4.0/).

Article

Simulation of Real Defect Geometry and Its Detection Using Passive Magnetic Inspection (PMI) Method

Milad Mosharafi [1,*], SeyedBijan Mahbaz [2] and Maurice B. Dusseault [2]

1. Mechanical and Mechatronics Engineering Department, University of Waterloo, Ontario, N2L 3G1, Canada
2. Earth and Environmental Sciences Department, University of Waterloo, Ontario, N2L 3G1, Canada; smahbaz@uwaterloo.ca (S.M.); mauriced@uwaterloo.ca (M.B.D.)
* Correspondence: mmoshara@uwaterloo.ca; Tel.: +1-519-504-3499

Received: 26 May 2018; Accepted: 4 July 2018; Published: 14 July 2018

Abstract: Reinforced concrete is the most commonly used material in urban, road, and industrial structures. Quantifying the condition of the reinforcing steel can help manage the human and financial risks that arise from unexpected reinforced concrete structure functional failure. Also, a quantitative time history of reinforcing steel condition can be used to make decisions on rehabilitation, decommissioning, or replacement. The self-magnetic behavior of ferromagnetic materials is useful for quantitative condition assessment. In this study, a ferromagnetic rebar with artificial defects was scanned by a three-dimensional (3D) laser scanner. The obtained point cloud was imported as a real geometry to a finite element software platform; its self-magnetic behavior was then simulated under the influence of Earth's magnetic field. The various passive magnetic parameters that can be measured were reviewed for different conditions. Statistical studies showed that 0.76% of the simulation-obtained data of the rebar surface was related to the defect locations. Additionally, acceptable coincidences were confirmed between the magnetic properties from numerical simulation and from experimental outputs, most noticeably at hole locations.

Keywords: reinforce concrete; rebar; defect; self-magnetic behavior; magnetic flux density; probability paper method; Passive Magnetic Inspection (PMI)

1. Introduction

Reinforced concrete as a composite infrastructure material is widely used in construction because of its excellent properties [1] and construction ease. Three factors control the behavioral responses of reinforced concrete: the reinforcing steel (generically referred to as rebar in this article), which has a noticeable ductile nature; the concrete itself, which has a noticeable brittle nature (low tensile strength but high compressive strength); and the condition of the rebar–concrete bonding (to achieve reliable stress transfer) [2].

Reinforced concrete is commonly used in infrastructure such as buildings, bridges, and highway construction [3]. The quality of a country's transportation system is mostly based on the conditions of its highway bridges, all of which contain steel. At the present time, apparently, approximately 28% of concrete bridge decks in the United States (US) and 33% of highway bridges in Canada can actually be considered operationally deficient or in a condition warranting the cessation of active service, mainly because of rebar corrosion [4].

Rebar corrosion is common in environmentally-exposed structures; it reduces the service life of these structures and impacts load-carrying capacity [5]. In the worst cases, structural failure occurs because corrosion reduces a stressed rebar's cross-sectional area [6] to the point of rupture. Rebar corrosion also degrades the bonding quality and can create cracks in the structure from volumetric expansion [7,8]. Bond deterioration leaves structures more vulnerable to vibrations related to daily usage or earthquakes [9].

The corrosion of steel rebar embedded in concrete falls into two categories: one is related to the specifications of the rebar and the concrete; the other includes the environmental conditions (temperature, humidity, pH, salinity, etc.) to which the structure is exposed [10]. Exposure to chloride ions, usually mostly from environmental exposure, is the most significant reason for rebar corrosion [11]. Long-term exposure to chloride ions deteriorates the passive layer of oxide on the steel rebar, eventually causing significant deterioration or structural failure, which can carry substantial economic loss [10]. To reduce safety threats and financial impact, corrosion-threatened rebar condition should be monitored so that risks can be quantitatively managed (repair, replace, restore) [12].

The visual inspection method (VI) is commonly used to assess the conditions of reinforced structures [13]. VI evaluates the external surface of the structure without directly assessing the internal conditions [14]. Even with detailed rubrics and photo imagery, VI methods are weak and semi-quantitative at best, and they must be done in conjunction with other non-destructive methods [15]. Reinforced concrete can be inspected for different types of defects using various types of non-destructive testing (NDT) methods [16]; the most common methods are potential measurement survey [17], galvanic current measurement [18], ground penetrating radar (GPR) [19], rebound hammer [20], ultrasonic [21], and radiography [22].

Each NDT method has limitations [23]; for instance, the macro-current measurement is complicated to interpret, since its results are influenced by the distance between anode and cathode and humidity [24]. GPR results are influenced by the existence of voids and variable internal moisture conditions [25], which can confound interpretations in many ways, such as confusion with background structures, shadowing, or false identification of gaps or previously repaired sites as being corrosion sites (Type I errors) [4]. Half-cell potential surveys can only mark corrosion locations; they give no information about the corrosion extent [26]. Ultrasonic pulse velocity (UPV) or Schmidt hammer techniques assess the mechanical properties of concrete with no information directly related to rebar corrosion [27]. Similarly, radiographic and acoustic inspections can assess concrete conditions, but give no direct information related to rebar conditions [28].

Some active magnetic-based methods such as magnetic flux leakage (MFL) can provide information directly related to the rebar corrosion condition [29]. Such methods need an external source such as electromagnets to properly magnetize objects during inspection [30], which increases assessment time and energy costs. These methods are challenging to perform on structures with complicated geometries [31], and complex rebar geometries can hamper clear interpretation of different data sets collected over time.

With the intent of providing a better measure that is quantitative and consistent, we introduce the passive magnetic method, which takes advantage of the Earth's natural magnetic field in order to inspect ferromagnetic structures [32]. Passive magnetic methods require no special preparation [33] or artificial magnetic source [34], and use anomalies in the passive magnetic flux density to locate defects [35]. This method can detect rebar defects such as corrosion sites or cracks [33], and stress changes that impact the crystalline ferromagnetic structure [36].

We built a Passive Magnetic Inspection (PMI) tool to exploit the passive magnetic concept and examine the corrosion condition of embedded rebar by scanning from the external concrete surface [8]. Preliminary successes have been described [37] in which solid rebar was sketched in COMSOL^R software version 5.3a (COMSOL Group, Stockholm, Sweden). based on a real rebar's geometry. It was then magnetized, assuming a certain value of magnetic field. Next, the passive magnetic behavior was investigated at a fixed distance from the rebar. Building on that work, in this current paper, the same ferromagnetic steel rebar with artificial defects is scanned with a three-dimensional (3D) laser scanner to generate a detailed point cloud of the structure. This point cloud then serves as the geometry basis for the finite element method software (COMSOL^R software), in studying how the Earth's magnetic field affects the rebar. Different magnetic properties of the object are extracted and interpreted at several distances from the rebar, and the parameters influencing them are investigated.

Additionally, a statistical detection method is presented as a new development in passive magnetic data processing and interpretation.

2. Theoretical Background and Methodology

The Earth's internal magnetic field is caused by liquid iron motions in the planetary core [38,39], plus contributions from other sources such as mantle movements, the nature of the lithosphere, etc. [40]. The magnetic field is a three-dimensional vector [41] with a harmonic pattern due to the globe's rotational movement [42]. The vector originates from the surface of the Earth and extends beyond the atmosphere, and its magnitude and orientation are functions of location [41] and time [40].

Natural magnetic fields and other influential local magnetic sources [37], combined with internal and external stresses, can change the scattered stray magnetic field of ferromagnetic materials [43]. Internal domain walls' displacement and magnetic-moment rotation in ferromagnetic materials happen under the influence of external magnetic fields [44], and there are relationships between the micro-magnetic characteristics of these materials and their mechanical responses [45]. For example, if the steel is deformed significantly in the presence of a magnetic field, the magnetization of the domains and their orientation within the steel are affected.

Self-magnetic flux leakage (SMFL) is assumed to take place in the stress concentration areas of ferromagnetic materials affected by mechanical load under the Earth's magnetic field [46], and this condition can remain even after removing the load, creating detectable magnetic leakage at the material surface [47]. Measuring SMFL at the surface of the materials helps in estimating their stress–strain states (SSSs), which is an important parameter in determining a structure's reliability [48]. Therefore, the relation between localized stress and oriented magnetic domains is useful for detecting defects in ferromagnetic materials within the background magnetic field of the Earth [49].

Magnetic field parameters at a point in space are represented by magnetic flux density (B) and an external magnetic field (H). B and H are vectors with a proportional magnitude and parallel directions. Magnetic flux density (B) represents the closeness of the magnetic field lines, and shows the strength of the magnetic field [50]. Also, Gauss's magnetic field law states that $\nabla B = 0$ [51]. H and B may have a complex relationship in magnetic materials [52], but engineers usually invoke the relation established by Faraday and Maxwell, which demonstrates that B is produced in a magnetizable material due to the existence of a primary magnetic field (H) [53].

Numerical simulation of the PMI method is performed based on the stray magnetic field (H_d) and the stray magnetic field energy (E_d) [37]. Hubert and Schäfer in 1998 [54] presented the relation for calculating the stray magnetic field (Equation (1)), based on summarizing Gauss's magnetic field law. In Equation (1), magnetic polarization (J) is the product of "volume-normalized magnetization" M, multiplied by "vacuum magnetic permeability of free space" μ_0. Additionally, a relation suggested for estimating the stray magnetic field energy uses the balance of the magnetic charges as well as their integration over the volume of the ferromagnetic material (Equation (2)).

$$div H_d = -div\left(\frac{J}{\mu_0}\right) \quad (1)$$

$$E_d = \frac{1}{2}\mu_0 \int_{all\ space} H_d^2 dV = -\frac{1}{2}\mu_0 \int_{sample} H_d \cdot J dV \quad (2)$$

Based on potential theory, volume charge density (λ_V)—Equation (3)—and surface charge density (σ_S)—equations (4) and (5)—are other parameters related to magnetization (M)—Equation (6)—and can be implemented for computing stray fields. Surface charge density is calculated by Equation (4) when there is just one magnetic medium; Equation (5) is applied when there are two varied different media with their own magnetization values and a specific vector perpendicular to the separation plane of those materials (n):

$$\lambda_V = -div M \quad (3)$$

$$\sigma_S = M \cdot n \tag{4}$$

$$\sigma_S = (M_1 - M_2) \cdot n \tag{5}$$

$$M(r) = J(r)/J_s \tag{6}$$

According to Equation (7), the stray field energy at a position (r) can be also calculated through the negative gradient of the potential of the stray field energy at a place ($\Phi_d(r)$) [55], where $\Phi_d(r)$—Equation (8)—is a function of magnetization saturation (J_s), volume charge density (λ_V), surface charge density (σ_S), and the derivative of the position vector (r'). Next, the magnetic field energy is obtained from Equation (9) through the integration functions of surface charge density and volume charge density over the volume and surface, respectively.

$$H_d(r) = -grad\Phi_d(r) \tag{7}$$

$$\Phi_d(r) = \frac{J_s}{4\pi\mu_0}\left[\int \frac{\lambda_V(r')}{|r-r'|}dV' + \int \frac{\sigma_S(r')}{|r-r'|}dS'\right] \tag{8}$$

$$E_d = J_s\left[\int \lambda_V(r)\Phi_d(r)dV + \int \sigma_S(r)\Phi_d(r)dS\right] \tag{9}$$

For conducting this research article, we scanned 373.87 mm of the surface of a ferromagnetic rebar (low-carbon steel), with a diameter of 16 mm, and two artificial defects (Table 1) [37], using a high-resolution 3D laser scanner (Figure 1a) [56]. The shape of the rebar was created with cloud points (Figure 1b) that were modified and converted to a mesh by Mesh Lab V1.3.2 (http://meshlab.sourceforge.net/). Subsequently, the produced mesh was imported to COMSOL^R software and converted to a discretized surface and solid, respectively (Figure 1c). The solid rebar was simulated via COMSOL^R software with regard to the magnetic field of the Earth, different components of magnetic flux density were investigated at different spacing, related simulation results were compared with our previous experimental results, and statistical approaches were introduced.

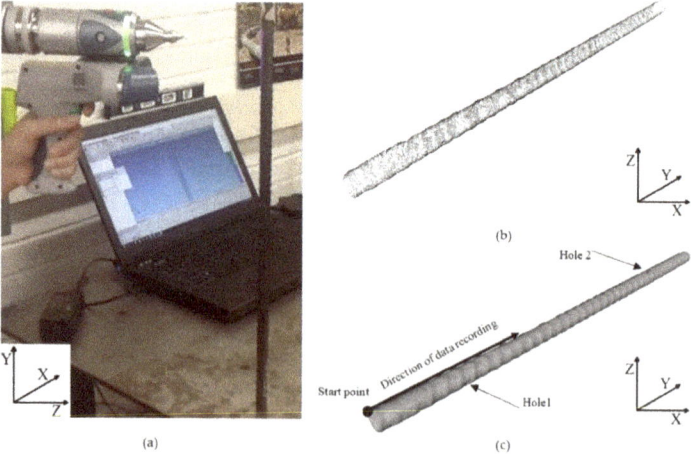

Figure 1. Process of converting the rebar geometry to a solid model: (**a**) scanning the rebar with three-dimensional (3D) laser scanner; (**b**) cloud points of rebar, presented in MeshLab; (**c**) solid illustration of rebar.

Table 1. Specifications of the two holes in the rebar.

Hole Name	Diameter (mm)	Depth (mm)	Y-Location from the Rebar's Start Point (mm)
Hole 1	0.58	1.24	57.91
Hole 2	0.68	0.57	282.67

3. Simulations and Results

After converting the rebar mesh to solid in COMSOL^R software, the magnetic behavior simulation was undertaken. Considering the variations of Earth's magnetic field in time and location, to obtain consistent and realistic results the average (within a year) of the different components of the magnetic field for the Waterloo, Ontario region (the location of the experiments) was adopted for the simulations (Table 2). Moreover, since the unitless relative magnetic permeability of low-carbon steels (ASTM 1020) range from 50 to 100 [57,58], a relative magnetic permeability of 75 was selected for this study.

Table 2. Background magnetic field (magnetic field of the Earth): from August 2016 to August 2017 (Adapted from Natural Resources Canada (http://www.nrcan.gc.ca)).

Background Magnetic Field (X-Component)	Background Magnetic Field (Y-Component)	Background Magnetic Field (Z-Component)
18 µT	−3 µT	50 µT

The duration of exposure to an external magnetic field will affect the magnetic behavior of ferromagnetic materials. In reality, ferromagnetic materials are affected by the magnetic field of the Earth from the beginning of their production process. There may also be some unknown external magnetic sources in the surrounding environment that affect the magnetic behavior of ferromagnetic objects [59]. However, as accurately as possible, we can apply the magnetic field of the Earth to the object and simulate its magnetic behavior, although some divergence will exist between the simulation and the experimental results.

To consider the Earth's magnetic field in the simulation, the rebar was located in a regular space (Figure 2) with dimensions of 100 mm × 150 mm × 410 mm, which included the magnetic field presented in Table 2 and Figure 3. To have better control of simulation parameters, the box and rebar were meshed separately with tetrahedral meshes according to the specifications in Table 3 (Figure 4a,b). Then, the rebar and box were jointly subjected to the simulation process as a single system (Figure 4c). The values of the different components (X, Y, and Z) of the magnetic flux densities were recorded for the Y direction of the rebar (i.e., the path parallel to the rebar's length). This path is at the surface of the rebar, and extends from one side (Edge A) to the other side of the box (Edge B) (Figure 5).

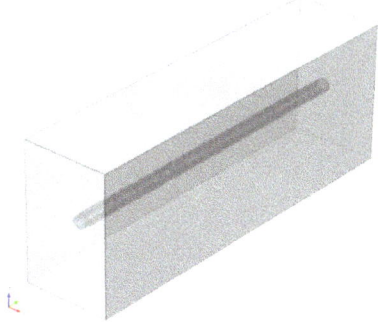

Figure 2. Solid rebar located in a box.

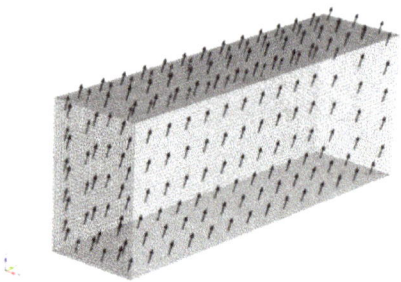

Figure 3. Box used in analysis; arrows show the resultant vector for X, Y, and Z components of Earth's magnetic field.

Table 3. Mesh specifications of rebar and box in the initial simulation.

Section Name	Rebar	Box
Maximum element size (mm)	2	8
Minimum element size (mm)	1	4.1
Maximum element growth rate	1.45	1.45
Curvature factor	0.5	0.5
Resolution of narrow regions	0.6	0.6
Number of degrees of freedom (in total)	601,773	

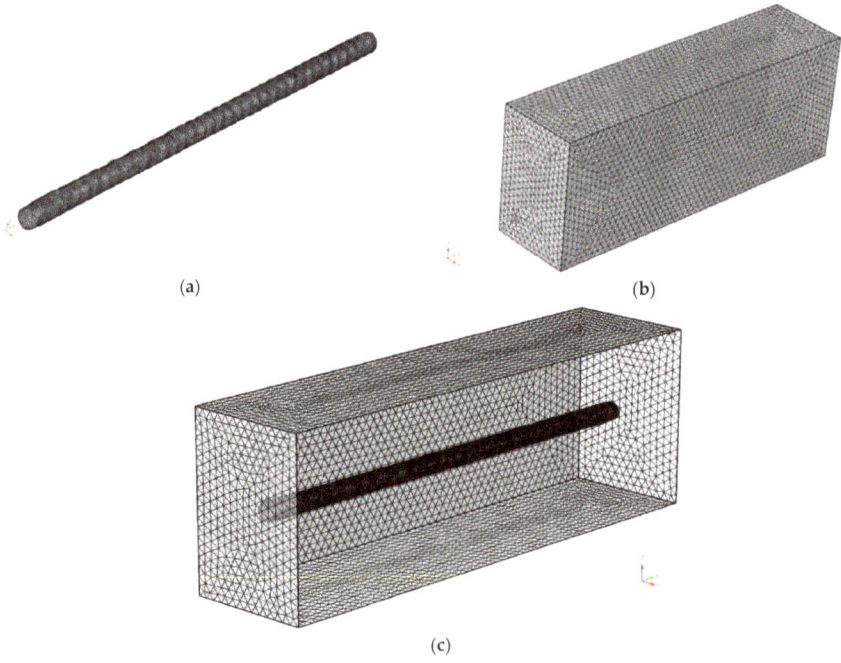

Figure 4. Initial meshes of the system: (**a**) rebar mesh with its initial sizes; (**b**) box mesh with its initial sizes; (**c**) rebar and box meshes as a single system (front face of the box is removed for better visualization).

As observed in Figure 6, at first, the values of all of the components of magnetic flux densities are equal to the background magnetic flux (the magnetic field of the Earth). When the Y distance reaches about 18.065 mm, at the end of the rebar, the values of all of the components begin to reflect the impact of the magnetic properties of the ferromagnetic rebar on the magnetic fluxes.

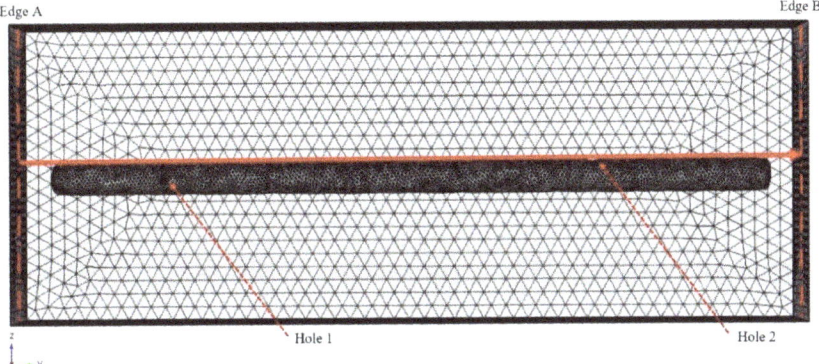

Figure 5. Path of the data recording (at the surface of the rebar in the Y direction).

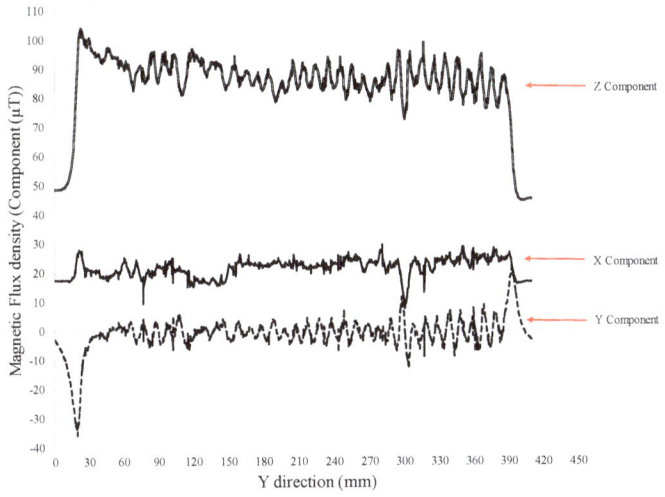

Figure 6. Values of different components (X, Y and Z) of the magnetic flux densities in the Y direction at the surface of the rebar (initial mesh of the rebar and box).

The values of all of the components have a harmonic variation because of the corrugated rebar shape. When the Y distance reaches the end of the rebar, all of the components of magnetic flux densities revert to the magnitudes of the background magnetic field. However, there is a distinguishable irregularity in the direction and values of all of the components at the location of Hole 2 (~301 mm from Edge A of the box). This irregularity is in the form of a minimum peak in the values of the Z and X magnetic flux densities, and in the form of a sudden change in the gradient of the Y-component of the magnetic flux density (a spike above the zero line, followed by a sudden dip below the zero line, then a sharp jump back to the zero line).

There are some outlier values in the different components of magnetic flux densities, which are related to the specifications of the elements used in this simulation. In order to have mesh element independent results, more accurate element specifications are implemented (Table 4). Then, the maximum and minimum values of the Z-component magnetic flux density (as a representative metric) from 295.0592 mm to 307.0592 mm (values symmetric about the fixed extent of Hole 2) are extracted. The location of Hole 2 was chosen because of its importance in our investigation. The differences between the maximum and minimum of these values are also used to verify the convergence of the simulation outcomes.

The values of the maximum and minimum magnetic flux densities become stable at mesh numbers 4 and 8 (Table 4), respectively (Figures 7 and 8). The difference between the maximum and minimum magnetic flux densities in the Z-component stabilize at rebar mesh #5 (Figure 9). Hence, the result of mesh #8 is used for continuing the simulation. The magnetic flux density values for mesh #8 have no out-of-range or disorder trend, compared with the trend of rebar mesh #1, which was the initial simulation (Figure 10).

Table 4. Different mesh specifications of rebar, with the fixed mesh specifications of box mesh as #1.

Mesh	Maximum Element Size (mm)	Minimum Element Size (mm)	Maximum Element Growth Rate	Curvature Factor	Resolution of Narrow Regions	Number of Degrees of Freedom (in Total)
1	2.000	1.000	1.450	0.500	0.600	601,773
2	1.340	0.670	1.407	0.450	0.636	1,267,526
3	0.898	0.449	1.364	0.405	0.674	3,324,359
4	0.602	0.301	1.323	0.365	0.715	9,764,894
5	0.571	0.286	1.310	0.361	0.722	10,441,703
6	0.5605	0.278	1.295	0.3505	0.746	10,995,911
7	0.550	0.270	1.280	0.340	0.770	11,594,725
8	0.530	0.240	1.260	0.330	0.780	12,877,797
9	0.500	0.200	1.250	0.320	0.790	15,173,763
10	0.460	0.160	1.220	0.280	0.810	19,243,609
11	0.446	0.141	1.100	0.240	0.830	20,879,674

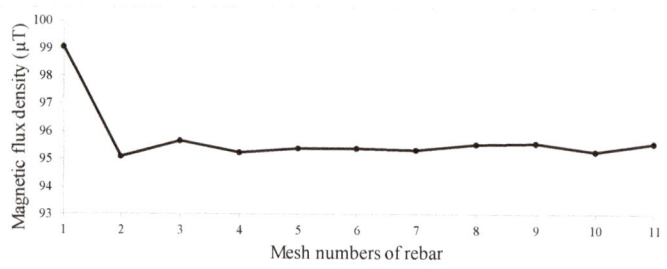

Figure 7. Maximum values of Z-component magnetic flux density, from 295.0592 mm to 307.0592 mm (values related to Hole 2), for different mesh specifications of rebar with fixed box mesh #1.

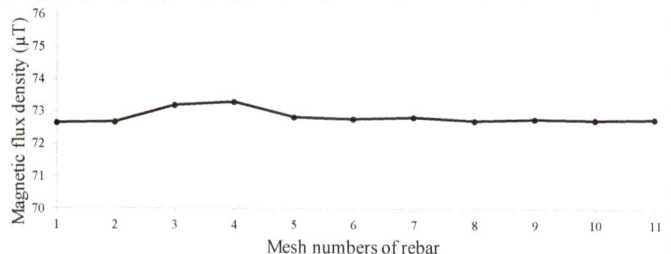

Figure 8. Minimum values of Z-component magnetic flux density, from 295.0592 mm to 307.0592 mm (values related to Hole 2), for different mesh specifications of rebar with fixed box mesh #1.

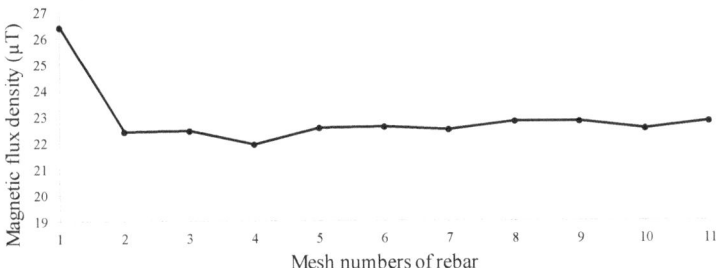

Figure 9. Difference between the maximum and minimum values of Z-component magnetic flux density, from 295.0592 mm to 307.0592 mm (values related to Hole 2), for different mesh specifications of rebar with fixed box mesh #1.

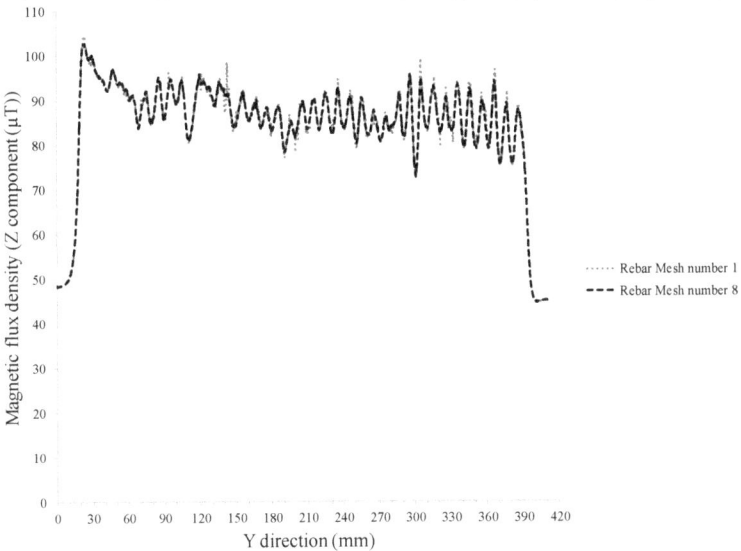

Figure 10. Comparison between the values of Z-component magnetic flux density of rebar mesh #8 and rebar mesh #1, with fixed box mesh (box mesh #1).

To determine the effect of spacing, values of the magnetic flux density (rebar mesh #8) with different spacing were investigated. It was understood that increasing the spacing between the rebar and the recording point would result in some outliers in the trend of the Z-component magnetic flux density, related to the specifications of the tetrahedral elements used in the box. To make the results of the simulation independent of the element specifications, more accurate element specifications were applied to the box (Table 5). As a representative result, the magnetic flux densities for the Z-component at a distance of 16 mm were extracted (Figure 11). The maximum and minimum values from 295.0592 mm to 307.0592 mm (values related to Hole 2) were reviewed. Subsequently, the difference between the maximum and minimum values was investigated, and we found that the values became stable with box mesh #5 (Figures 12–14).

Table 5. Different mesh specifications of box, with the fixed mesh specifications of rebar (rebar mesh #8).

Mesh	Maximum Element Size (mm)	Minimum Element Size (mm)	Maximum Element Growth Rate	Curvature Factor	Resolution of Narrow Regions	Number of Degrees of Freedom (in Total)
1	8.000	4.100	1.450	0.500	0.600	12,877,797
2	7.720	3.400	1.330	0.410	0.620	13,794,957
3	6.820	2.300	1.300	0.400	0.650	14,188,984
4	5.810	1.400	1.250	0.350	0.680	15,058,001
5	4.110	1.100	1.190	0.290	0.710	17,446,126
6	2.840	0.850	1.150	0.250	0.730	22,627,445
7	2.250	0.820	1.140	0.230	0.730	28,481,960
8	2.210	0.815	1.130	0.230	0.740	29,650,862

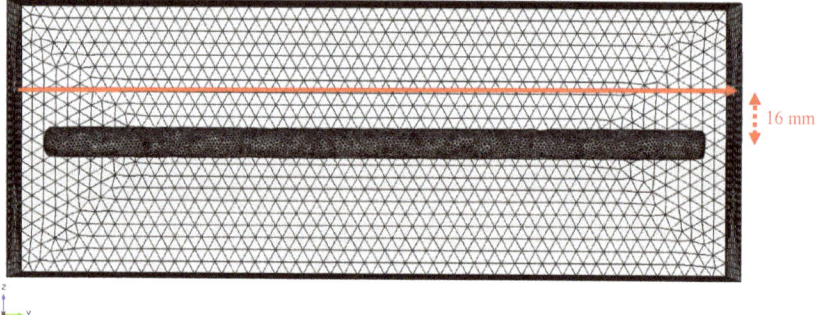

Figure 11. Path of data recording (with distance 16 mm from center of the rebar).

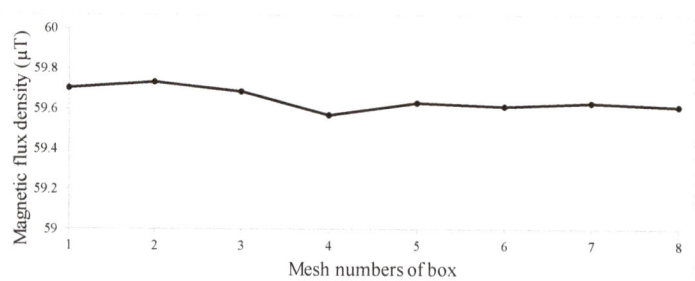

Figure 12. Maximum values of Z-component magnetic flux density, from 295.0592 mm to 307.0592 mm (values related to Hole 2), for different box mesh specifications with fixed rebar mesh #8 (Table 4).

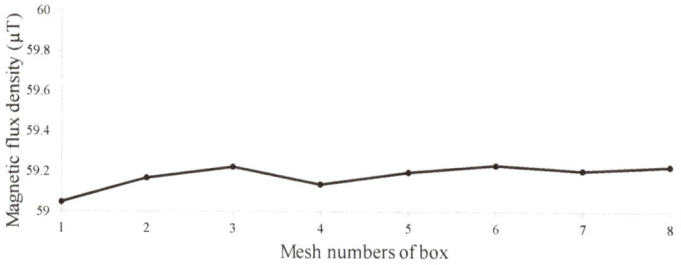

Figure 13. Minimum values of Z-component magnetic flux density, from 295.0592 mm to 307.0592 mm (values related to Hole 2), for different box mesh specifications with a fixed rebar mesh #8 (Table 4).

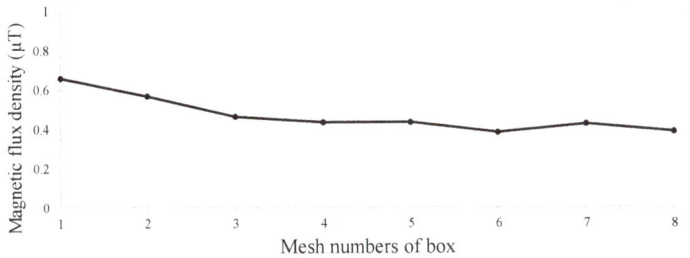

Figure 14. Difference between the maximum and minimum values of Z-component magnetic flux density, from 295.0592 mm to 307.0592 mm (values related to Hole 2), for different box mesh specifications with fixed rebar mesh #8 (Table 4).

According to Figures 7–9 and 12–14, the outcomes from the simulation of the rebar with mesh #8 and box mesh #5, the optimum mesh specifications, were chosen for the rest of the investigations. After the optimum mesh specifications, achieved by increasing the meshing accuracy, the biggest difference in the maximum and minimum values of magnetic flux densities (of the same place) are respectively less than 0.01 µT and 0.04 µT, which are considered negligible. Additionally, the same values are observed for the maximum and minimum values of magnetic data when the element accuracy is increased. The same magnetic values mean that the results converge and are independent of the mesh specifications.

Carrying out simulations with optimum mesh specifications led to a graphical representation (Figure 15), which shows the behavior of the Z-component magnetic flux density at the location of Hole 2. Also, a planar slice of the magnetic field under the rebar in Figure 15 shows the conditions of the stray magnetic field around the rebar. As the distance from the rebar increases, the stray magnetic field around the rebar decreases relatively uniformly and symmetrically.

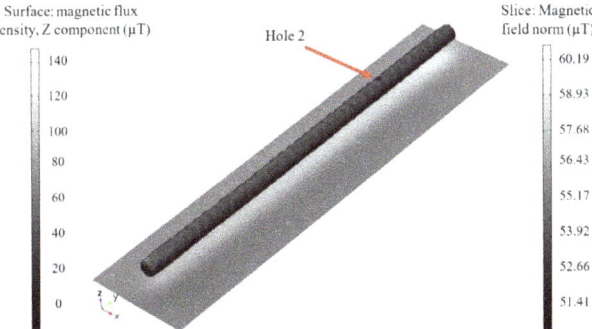

Figure 15. Behavior of Z-component magnetic flux density and normal magnetic field around the rebar (rebar mesh #8 and box mesh #5).

Figure 16 shows the values of magnetic flux densities of rebar with optimum mesh specifications at different spacings from the center of the rebar, ranging from 8 mm (surface level) to 72 mm (maximum distance from the surface). The behavior of the Z-component magnetic flux density is distinguishable at Hole 2 at a maximum of 16 mm from the rebar's center (Figure 16), which is a distance equal to 8 mm from the rebar's surface (this distance represents how thick the concrete above the bar can be for detection of the buried defects). According to the simulation results, it seems that the technique can be used only for very thin concrete layers with a maximum thickness of 8 mm.

It should be mentioned that the simulations were performed under the Earth's present magnetic field, but ferromagnetic materials are considered saturated by the natural magnetic field, and may show stronger magnetic behavior.

For further investigation, the data-recording distance was increased to the maximum possible distance from the rebar, aligning with the inside edge of the box. At larger distances, the magnetic flux density trend becomes smoother and straighter, and approaches the background magnetic field. The minimum values of the Z-component magnetic flux density, from 295.0592 mm to 307.0592 mm (values related to Hole 2), were considered for different distances. Increasing the vertical distance (in the Z direction) of the data recording line logarithmically decreased the minimum value of the Z-component magnetic flux density until this value reached an approximately constant value. The trend line showing the relation between the minimum values of the Z-component magnetic flux density and data recorded at various distances is a fourth-order polynomial equation (Figure 17).

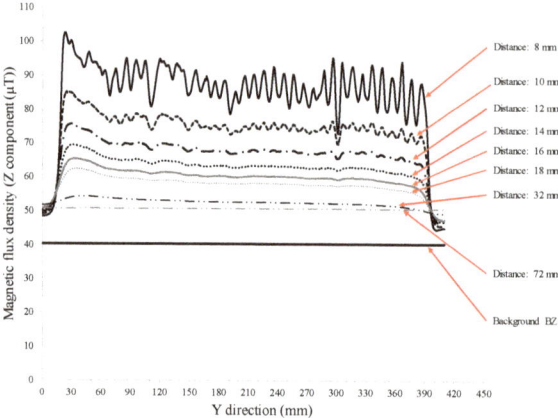

Figure 16. Values of magnetic flux densities of rebar mesh #8 and box mesh #5 at different vertical distances from the center of the rebar.

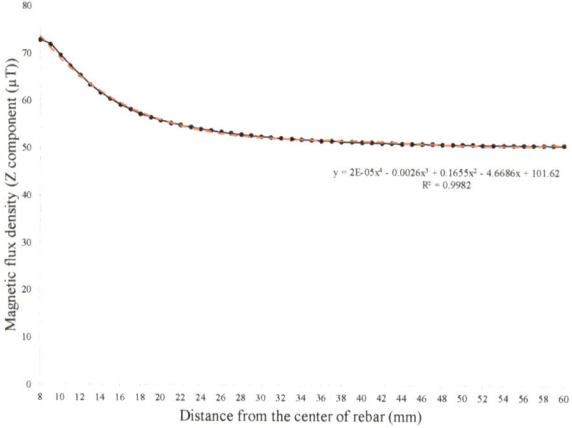

Figure 17. Behavior of the minimum values of the Z-component magnetic flux density, from 295.0592 mm to 307.0592 mm (values related to Hole 2), of rebar mesh #8 and box mesh #5, recorded at different vertical distances.

We reviewed the different components of magnetic flux densities at the surface of the rebar, which were extracted from the optimum mesh specifications (Figure 18). The noise and out-of-range values at their minimum and results correlate well with the experimental results reported previously (Figure 19) [37]. The laboratory flux magnetic density measurements were conducted using the PMI device that was specifically developed for this work in our lab, and is on its way to being commercialized. The PMI device works through scanning the SMFL arising from ferromagnetic structures [37].

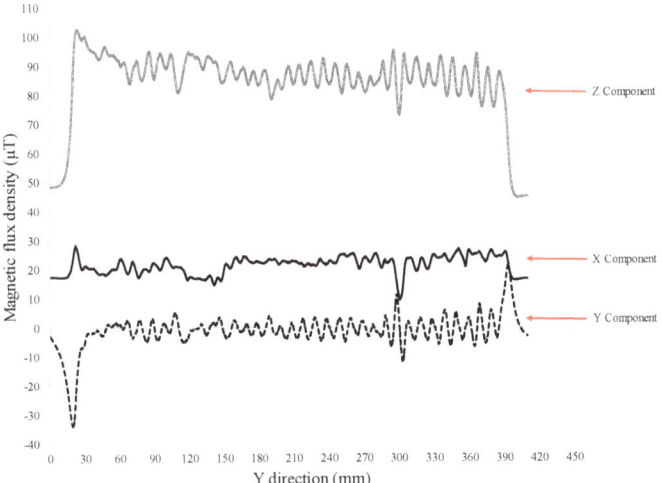

Figure 18. Magnetic flux density values at different axes (X, Y and Z) in the Y direction at the surface of the rebar (rebar mesh #8 and box mesh #5).

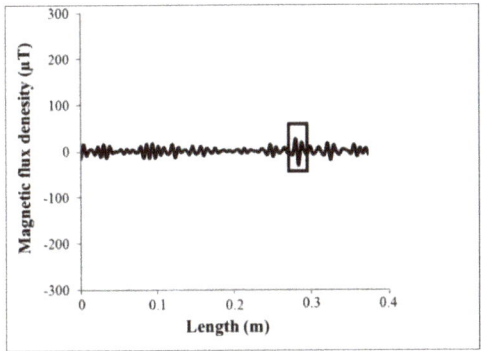

Figure 19. X-component of magnetic flux density resulting from the previous experiments; the square shows the Hole 2 location (modified from Mahbaz et al., 2017 [37]).

The patterns of laboratory and simulated outputs at the holes' locations follow the same trend; both curves generally have an up and down trend due to the corrugated shape of the rebar, along with a minimum value at the center of the Hole 2. As seen in Figure 18, the top hole that is ~301 mm from Edge A of the box (equal to ~282 mm from the rebar's start point (Figure 19)) is substantially easier to detect than the hole on the side of the rebar. Finding the irregularity in the magnetic data related to Hole 2 is easier because of the difference between the magnetic property of its surface (filled with air) and the rest of the rebar's surface (which has a different magnetic property). No detectable irregularity can be sensed in the surface magnetic flux densities for Hole 1, because more metal lies between it and the scanning line.

4. Discussion

Assuming that the magnetic flux densities of different locations on the rebar are independent of one another, the probability graph method was used for fitting the magnetic flux values to a probability distribution. This method involves equating the empirical distribution of magnetic data (P_i) with the chosen cumulative distribution function (CDF). Next, the CDF function is written in linear form, which can be expressed as Equation (10) for Gamma CDF (GAMMADIS). The linearity is then used as a basis for determining whether the data can be modeled by a particular distribution. Additionally, the goodness-of-fit of the model is given by the coefficient of determination R^2.

$$P_i = GAMMADIS(magnetic\ data) \rightarrow magnetic\ data = GAMMADIS_{inverse}(P_i) \qquad (10)$$

The magnetic flux density data were plotted against various probability distributions (normal, log-normal, Weibull, and gamma distributions); a gamma distribution was chosen based on the minimum least-squared error (Figure 20). This distribution is based on a function of two parameters: α and β (Equation (11)); these were calculated by the mean and standard deviations (SD), which are 87.8 µT and 25.6 µT, respectively. As observed in Figure 21, the gamma function correlates well with the histogram frequency of data, and this approximation may be useful for estimation in practical cases.

$$f(x) = \frac{1}{\beta^\alpha G(\alpha)} x^{\alpha-1} e^{-\frac{x}{\beta}} \qquad (11)$$

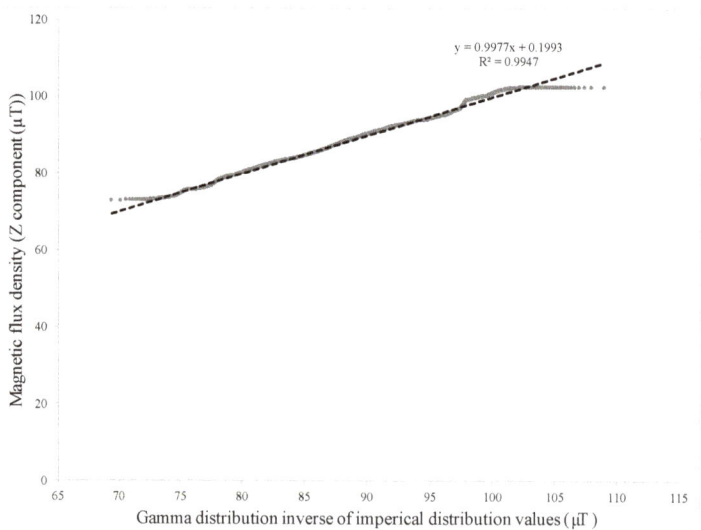

Figure 20. Probability plot for investigating the correlation of data with a gamma distribution.

According to Figure 18, a Z-component magnetic flux density of less than 76 µT (without considering the edge effect and background magnetic field) corresponds to the location of Hole 2. Importing this value into the obtained CDF shows that 0.76% of the data is related to the defective locations. In other words, 0.76% of the rebar surface (at the scanned section) can be considered imperfect. This result can be verified by the Monte Carlo simulation method (based on inverse values of the obtained gamma distribution function). Figure 22 presents the probability of defects considering the mean, SD, and limit state, showing that the probability of defectiveness fluctuates until the first 300 trails are completed, and then stabilizes at the value of ~0.75%.

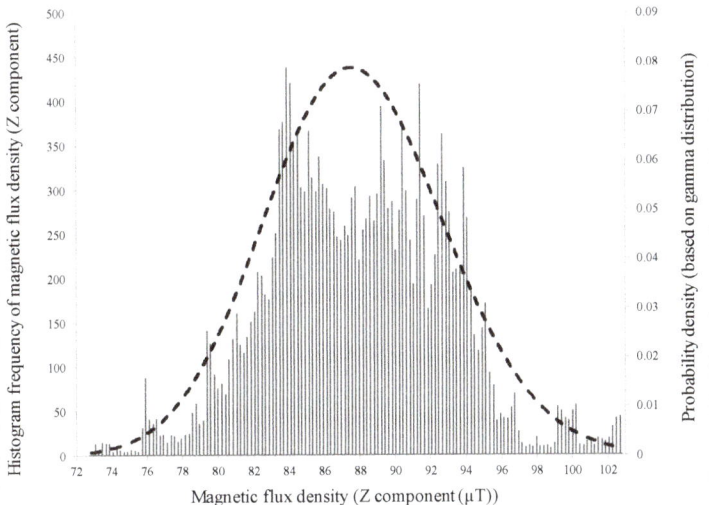

Figure 21. Histogram frequency of data in conjunction with gamma distribution probability density.

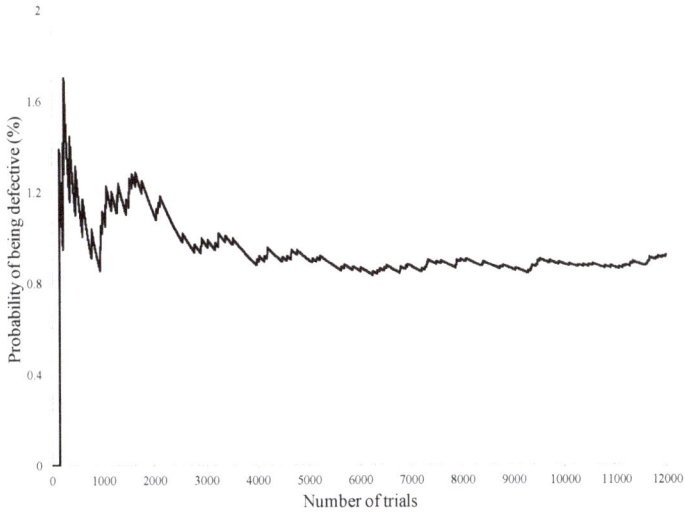

Figure 22. Defectiveness probability for the inspected rebar based on the Monte Carlo simulation method.

For our statistical investigations, we considered the magnetic data as independent variables. Those independent variables were described by the chosen probability distribution with its particular distribution parameters, knowing that distribution allowed us to estimate an interval over which the unknown future values may lie (with a stipulated level of confidence). Using the CDF of the gamma distribution, about 98% of all of the data are from 76 µT to 100 µT (Equation (12)). Hence, regarding the recorded magnetic data of the rebar, it can be predicted with 98% confidence that if the rebar was longer (by how much is irrelevant), the next values indicating a flawless rebar would be somewhere between 76–100 µT. Values outside this range should be reviewed as suspected defect locations.

$$GAMMADIS(100\ \mu T) - GAMMADIS(76\ \mu T) = 0.99 - 0.01 = 0.98 \tag{12}$$

Figure 23 shows the values of different components of magnetic flux densities, 16 mm from the center of the rebar. For better observation of the irregularities related to Hole 2, all of the graphs are presented from 200 mm to 370 mm at the appropriate scale. The anomaly in the magnetic values at the location of Hole 2 can be detected in all of the components of the magnetic flux densities. Figure 23 indicates that the values of the X and Y-components of magnetic flux density still include some noise that is attributable to the mesh specifications. This issue can be investigated by increasing the mesh density in both the rebar and the box containing the rebar. Additionally, increasing the quality of clouds points defining the bar geometry can help achieve more accurate outcomes. For instance, Hole 1 in the solid part produced from the captured cloud points was not as deep as the real depth (measured directly), but the mesh was not corrected, as the intent was in part to test the laser scan cloud point data without additional data input management.

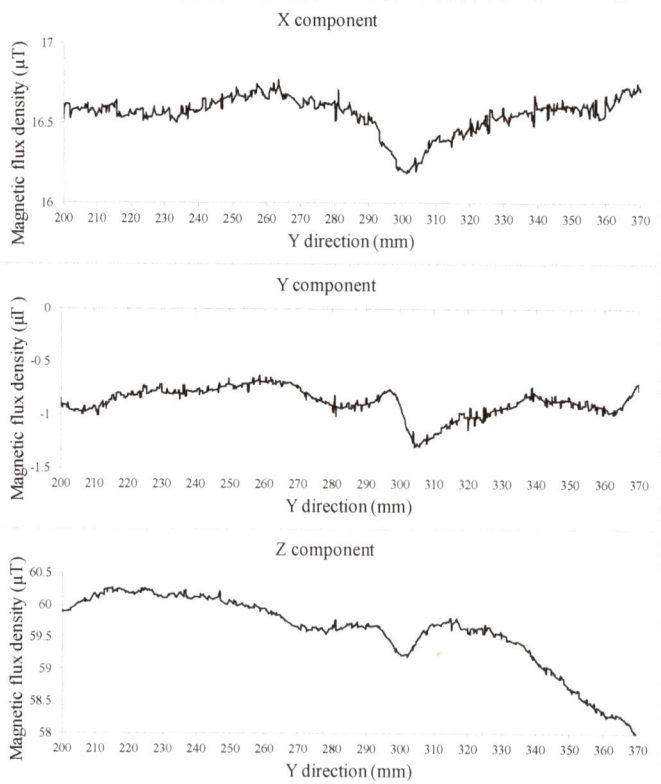

Figure 23. Values of different components (X, Y, and Z) of the magnetic flux densities in the Y direction, 16 mm from the center of the rebar (rebar mesh #8 and box mesh #5).

5. Conclusions

Being able to detect defects in steel infrastructure would substantially improve risk management and condition evaluation over time. To this end, mathematical simulations were carried out on a pre-flawed specimen that was laser-scanned to generate a point cloud surface map. This map was then used as a basis to develop a model. The intent was to establish detectability limits for very small flaws in order to reduce Type I and Type II errors in anomaly detection.

The magnetic behavior of the ferromagnetic rebar specimen was simulated with a finite element-based software considering the background magnetic field. Different components of magnetic flux densities on the surface showed consistent harmonic trends because of the corrugated shape of the rebar. However, there were specific irregularities in the direction and values for the different components of magnetic flux densities at the location of Hole 2. Simulated patterns could be correlated with the experimental data at the holes' locations, so the top hole (Hole 2) was easily located, but Hole 1 was not, because of its orientation in the magnetic field and because the point cloud model did not replicate its true depth. The gamma probability distribution was chosen to statistically assess the magnetic flux density behavior of the rebar. Two main outcomes were extracted: 0.76% of the scanned section of the rebar was considered defective, and if the rebar specimen were longer, the Z-component magnetic flux density values indicating flawless rebar would be predicted to lie between 76–100 µT with 98% confidence.

The values of the different components of magnetic flux densities at different distances from the rebar were reviewed. Increasing the vertical distance of the data recording line led to a logarithmic reduction of magnetic flux density values. As this distance was increased, the magnetic flux density values became approximately constant and close to the background magnetic field. In conclusion:

- The pattern of the simulation results at defect locations were similar to the outputs of previous physical experiments;
- The background magnetic field had a significant effect on the trend and values of different components of the magnetic flux density;
- All of the magnetic flux density components displayed correctly located anomalies corresponding to the defect on the top surface of the rebar;
- Increasing the distance from the rebar changed the trend and values of the magnetic flux densities such that at some distance, the anomaly became undetectable;
- To detect various shapes and sizes of defects at different places along a rebar specimen, additional magnetic parameters should be considered. For instance, the Z-component of the magnetic flux density was totally constant on the sides of the rebar, and could not detect the anomaly arising from Hole 1;
- The stray magnetic field around the rebar decreased relatively symmetrically by increasing the distance from the rebar; and
- The choice of the gamma distribution to model the Z-component magnetic flux density values of the numerical simulation resulted in valuable interpretations.

Author Contributions: Conceptualization, M.M., S.M. and M.B.D.; Data curation, M.M. and S.M.; Formal analysis, M.M.; Funding acquisition, M.B.D.; Investigation, M.M.; Methodology, S.M.; Supervision, M.B.D.; Writing—original draft, M.M.; Writing—review & editing, M.M., S.M., and M.B.D.

Funding: This research received no direct funding.

Acknowledgments: The authors would like to thank Inspecterra Inc. for providing access to the PMI device used in this study. We would like to acknowledge CMC Microsystems for the provision of products and services that facilitated this research, including the providing of a COMSOL^R software license. In addition, we express our thanks to Professor Ralph Haas and Professor Carl T. Haas for permitting us to use their laboratory (Infrastructure and sensing analysis laboratory) during our studies. we also extend our thanks to their lab member Mr. Mohammad-Mahdi Sharif due to his support for scanning the rebar by the 3D-laser scanner.

Conflicts of Interest: The authors declare no conflict of interest.

References

1. Chandramauli, A.; Bahuguna, A.; Javaid, A. The analysis of plain cement concrete for future scope when mixed with glass & fibres. *IJCIET* **2018**, *9*, 230–237.
2. Hameed, R.; Sellier, A.; Turatsinze, A.; Duprat, F. Simplified approach to model steel rebar-concrete interface in reinforced concrete. *KSCE J. Civ. Eng.* **2017**, *21*, 1291–1298. [CrossRef]
3. Boyle, H.C.; Karbhari, V.M. Bond and behavior of composite reinforcing bars in concrete. *Polym. Plast. Technol. Eng.* **1995**, *34*, 697–720. [CrossRef]
4. Abouhamad, M.; Dawood, T.; Jabri, A.; Alsharqawi, M.; Zayed, T. Corrosiveness mapping of bridge decks using image-based analysis of GPR data. *Autom. Constr.* **2017**, *80*, 104–117. [CrossRef]
5. Li, F.; Ye, W. A Parameter Sensitivity Analysis of the Effect of Rebar Corrosion on the Stress Field in the Surrounding Concrete. *Adv. Mater. Sci. Eng.* **2017**, *2017*, 9858506. [CrossRef]
6. Peng, J.; Tang, H.; Zhang, J. Structural Behavior of Corroded Reinforced Concrete Beams Strengthened with Steel Plate. *J. Perform. Constr. Facil.* **2017**, *31*, 04017013. [CrossRef]
7. Desnerck, P.; Lees, J.M.; Morley, C.T. Bond behaviour of reinforcing bars in cracked concrete. *Constr. Build. Mater.* **2015**, *94*, 126–136. [CrossRef]
8. Mahbaz, S.B. Non-Destructive Passive Magnetic and Ultrasonic Inspection Methods for Condition Assessment of Reinforced Concrete. Ph.D. Thesis, Department of Civil and Environmental Engineering, University of Waterloo, Waterloo, ON, Canada, 2016.
9. Shi, Y.; Li, Z.X.; Hao, H. Bond slip modelling and its effect on numerical analysis of blast-induced responses of RC columns. *Struct. Eng. Mech.* **2009**, *32*, 251–267. [CrossRef]
10. Valipour, M.; Shekarchi, M.; Ghods, P. Comparative studies of experimental and numerical techniques in measurement of corrosion rate and time-to-corrosion-initiation of rebar in concrete in marine environments. *Cem. Concr. Compos.* **2014**, *48*, 98–107. [CrossRef]
11. Montemor, M.F.; Simões, A.M.P.; Ferreira, M.G.S. Chloride-induced corrosion on reinforcing steel: From the fundamentals to the monitoring techniques. *Cem. Concr. Compos.* **2003**, *25*, 491–502. [CrossRef]
12. Muchaidze, I.; Pommerenke, D.; Chen, G. Steel reinforcement corrosion detection with coaxial cable sensors. *Proc. SPIE Int. Soc. Opt. Eng.* **2011**, *7981*, 79811L.
13. Farhidzadeh, A.; Ebrahimkhanlou, A.; Salamone, S. A vision-based technique for damage assessment of reinforced concrete structures. *Proc. SPIE Int. Soc. Opt. Eng.* **2014**, *9064*, 90642H.
14. Takahashi, K.; Okamura, S.; Sato, M. A fundamental study of polarimetric gb-sar for nondestructive inspection of internal damage in concrete walls. *Electron. Commun. Jpn.* **2015**, *98*, 41–49. [CrossRef]
15. Concu, G.; de Nicolo, B.; Pani, L. Non-destructive testing as a tool in reinforced concrete buildings refurbishments. *Struct. Surv.* **2011**, *29*, 147–161. [CrossRef]
16. Szymanik, B.; Frankowski, P.K.; Chady, T.; Chelliah, C.R.A.J. Detection and inspection of steel bars in reinforced concrete structures using active infrared thermography with microwave excitation and eddy current sensors. *Sensors* **2016**, *16*, 234. [CrossRef] [PubMed]
17. Schneck, U. *Concrete Solutions 2014*; CRC Press: Boca Raton, FL, USA, 2014; pp. 577–585, ISBN 9781138027084.
18. Hardon, R.G.; Lambert, P.; Page, C.L. Relationship between electrochemical noise and corrosion rate of steel in salt contaminated concrete. *Br. Corros. J.* **1998**, *23*, 225–228. [CrossRef]
19. Kaur, P.; Dana, K.J.; Romero, F.A.; Gucunski, N. Automated GPR rebar analysis for robotic bridge deck evaluation. *IEEE Trans. Cybern.* **2016**, *46*, 2265–2276. [CrossRef] [PubMed]
20. Sanchez, J.; Andrade, C.; Torres, J.; Rebolledo, N.; Fullea, J. Determination of reinforced concrete durability with on-site resistivity measurements. *Mater. Struct.* **2017**, *50*, 41. [CrossRef]
21. Sabbağ, N.; Uyanık, O. Prediction of reinforced concrete strength by ultrasonic velocities. *J. Appl. Geophys.* **2017**, *141*, 13–23. [CrossRef]
22. Pei, C.; Wu, W.; Ueaska, M. Image enhancement for on-site X-ray nondestructive inspection of reinforced concrete structures. *J. X-ray Sci. Technol.* **2016**, *24*, 797–805. [CrossRef] [PubMed]
23. Hussain, A.; Akhtar, S. Review of non-destructive tests for evaluation of historic masonry and concrete structures. *Arab. J. Sci. Eng.* **2017**, *42*, 925–940. [CrossRef]
24. Xu, C.; Li, Z.; Jin, W. A new corrosion sensor to determine the start and development of embedded rebar corrosion process at coastal concrete. *Sensors* **2013**, *13*, 13258–13275. [CrossRef] [PubMed]

25. Evans, R.D.; Rahman, M. *Advances in Transportation Geotechnics 2*; CRC Press: Boca Raton, FL, USA, 2012; pp. 516–521, ISBN 9780415621359.
26. Owusu Twumasi, J.; Le, V.; Tang, Q.; Yu, T. Quantitative sensing of corroded steel rebar embedded in cement mortar specimens using ultrasonic testing. *Proc. SPIE Int. Soc. Opt. Eng.* **2016**, *9804*, 98040P.
27. Verma, S.K.; Bhadauria, S.S.; Akhtar, S. Review of nondestructive testing methods for condition monitoring of concrete structures. *Can. J. Civ. Eng.* **2013**, *2013*, 834572. [CrossRef]
28. Perin, D.; Göktepe, M. Inspection of rebars in concrete blocks. *Int. J. Appl. Electromagn. Mech.* **2012**, *38*, 65–78.
29. Makar, J.; Desnoyers, R. Magnetic field techniques for the inspection of steel under concrete cover. *NDT E Int.* **2001**, *34*, 445–456. [CrossRef]
30. Daniel, J.; Abudhahir, A.; Paulin, J. Magnetic flux leakage (MFL) based defect characterization of steam generator tubes using artificial neural networks. *J. Magn.* **2017**, *22*, 34–42. [CrossRef]
31. Wang, Z.D.; Gu, Y.; Wang, Y.S. A review of three magnetic NDT technologies. *J. Magn. Magn. Mater.* **2012**, *324*, 382–388. [CrossRef]
32. Doubov, A. Screening of weld quality using the magnetic metal memory effect. *Weld World* **1998**, *41*, 196–199.
33. Ahmad, M.I.M.; Arifin, A.; Abdullah, S.; Jusoh, W.Z.W.; Singh, S.S.K. Fatigue crack effect on magnetic flux leakage for A283 grade C steel. *Steel Compos. Struct.* **2015**, *19*, 1549–1560. [CrossRef]
34. Gontarz, S.; Mączak, J.; Szulim, P. Online monitoring of steel constructions using passive methods. In *Engineering Asset Management—Systems, Professional Practices and Certification*; Lecture Notes in Mechanical Engineering; Springer: Cham, Switzerland, 2015; pp. 625–635.
35. Miya, K. Recent advancement of electromagnetic nondestructive inspection technology in Japan. *IEEE Trans. Magn.* **2002**, *38*, 321–326. [CrossRef]
36. Witos, M.; Zieja, M.; Zokowski, M.; Roskosz, M. Diagnosis of supporting structures of HV lines with using of the passive magnetic observer. *JSAEM Stud. Appl. Electromagn. Mech.* **2014**, *39*, 199–206.
37. Mahbaz, S.B.; Dusseault, M.B.; Cascante, G.; Vanheeghe, Ph. Detecting defects in steel reinforcement using the passive magnetic inspection method. *J. Environ. Eng. Geophys.* **2017**, *22*, 153–166. [CrossRef]
38. Hughes, D.W.; Cattaneo, F. Strong-field dynamo action in rapidly rotating convection with no inertia. *Phys. Rev. E* **2016**, *53*, 6200108. [CrossRef] [PubMed]
39. Davies, C.; Constable, C. Geomagnetic spikes on the core-mantle boundary. *Nat. Commun.* **2017**, *8*, 15593. [CrossRef] [PubMed]
40. Bezděk, A.; Sebera, J.; Klokočník, J. Validation of swarm accelerometer data by modelled nongravitational forces. *Adv. Space Res.* **2017**, *59*, 2512–2521. [CrossRef]
41. Taylor, B.K.; Johnsen, S.; Lohmann, K.J. Detection of magnetic field properties using distributed sensing: A computational neuroscience approach. *Bioinspir. Biomim.* **2017**, *12*, 036013. [CrossRef] [PubMed]
42. Zagorski, P.; Bangert, P.; Gallina, A. Identification of the orbit semi-major axis using frequency properties of onboard magnetic field measurements. *Aerosp. Sci. Technol.* **2017**, *66*, 380–391. [CrossRef]
43. Mironov, S.; Devizorova, Z.; Clergerie, A.; Buzdin, A. Magnetic mapping of defects in type-II superconductors. *Appl. Phys. Lett.* **2016**, *108*, 212602. [CrossRef]
44. Guo, L.; Shu, D.; Yin, L.; Chen, J.; Qi, X. The effect of temperature on the average volume of Barkhausen jump on Q235 carbon steel. *J. Magn. Magn. Mater.* **2016**, *407*, 262–265. [CrossRef]
45. Gupta, B.; Szielasko, K. Magnetic Sensor Principle for Susceptibility Imaging of Para- and Diamagnetic Materials. *J. Nondestruct. Eval.* **2016**, *35*, 41. [CrossRef]
46. Huang, H.; Qian, Z. Effect of temperature and stress on residual magnetic signals in ferromagnetic structural steel. *IEEE Trans. Magn.* **2017**, *53*, 6200108. [CrossRef]
47. Yuan, J.; Zhang, W. Detection of stress concentration and early plastic deformation by monitoring surface weak magnetic field change. In Proceedings of the IEEE International Conference on Mechatronics and Automation, Xi'an, China, 4–7 August 2010; pp. 395–400.
48. Dubov, A.A. Development of a metal magnetic memory method. *Chem. Pet. Eng.* **2012**, *47*, 837–839. [CrossRef]
49. Jarram, P. Remote measurement of stress in carbon steel pipelines—Developments in remote magnetic monitoring. In *NACE International Corrosion Conference Proceedings*; NACE International: Houston, TX, USA, 2016; pp. 1–9.
50. Tauxe, L. *Essentials of Paleomagnetism*; University of California Press: Berkeley, CA, USA, 2010; pp. 1–4, ISBN 0520260317, 9780520260313.

51. Hu, K.; Ma, Y.; Xu, J. Stable finite element methods preserving ∇ · B = 0 exactly for MHD models. *Numer. Math.* **2017**, *135*, 371–396. [CrossRef]
52. Tabrizi, M. The nonlinear magnetic core model used in spice plus. In Proceedings of the Applied Power Electronics Conference and Exposition, San Diego, CA, USA, 2–6 March 1987; pp. 32–36.
53. Tanel, Z.; Erol, M. Students' difficulties in understanding the concepts of magnetic field strength, magnetic flux density and magnetization. *Lat.-Am. J. Phys. Educ.* **2008**, *2*, 184–191.
54. Hubert, A.; Schafer, R. *Magnetic Domains: The Analysis of Magnetic Microstructures*; Springer: Berlin/Heidelberg, Germany, 1998; pp. 109–110. ISBN 978-3-540-64108-7.
55. Kronmuller, H. Theory of nucleation fields in inhomogeneous ferromagnets. *Phys. Status Solidi B* **1987**, *144*, 385–396. [CrossRef]
56. Nahangi, M.; Haas, C. Automated 3D compliance checking in pipe spool fabrication. *Adv. Eng. Inform.* **2014**, *28*, 360–369. [CrossRef]
57. Rose, J.H.; Uzal, E.; Moulder, J.C. *Magnetic Permeability and Eddy Current Measurements*; Springer: Boston, MA, USA, 1995; Volume 14, pp. 315–322.
58. Ribichini, R. Modelling of Electromagnetic Acoustic Transducers. Ph.D. Thesis, Department of Mechanical Engineering, Imperial College, London, UK, 2011.
59. Li, Z.; Jarvis, R.; Nagy, P.B.; Dixon, S.; Cawley, P. Experimental and simulation methods to study the Magnetic Tomography Method (MTM) for pipe defect detection. *NDT E Int.* **2017**, *92*, 59–66. [CrossRef]

© 2018 by the authors. Licensee MDPI, Basel, Switzerland. This article is an open access article distributed under the terms and conditions of the Creative Commons Attribution (CC BY) license (http://creativecommons.org/licenses/by/4.0/).

Article

Defect Detection of Aluminum Alloy Wheels in Radiography Images Using Adaptive Threshold and Morphological Reconstruction

Junsheng Zhang, Zhijie Guo, Tengyun Jiao and Mingquan Wang *

Science and Technology on Electronic Test and Measurement Laboratory, North University of China, Taiyuan 030051, China; zhangsheng658@163.com (J.Z.); 18734551657@163.com (Z.G.); 18834160280@163.com (T.J.)
* Correspondence: wangmq@nuc.edu.cn

Received: 5 November 2018; Accepted: 20 November 2018; Published: 23 November 2018

Abstract: In low-pressure casting, aluminum alloy wheels are prone to internal defects such as gas holes and shrinkage cavities, which call for X-ray inspection to ensure quality. Automatic defect segmentation of X-ray images is an important task in X-ray inspection of wheels. For this, a solution is proposed here that combines adaptive threshold segmentation algorithm and mathematical morphology reconstruction. First, the X-ray image of the wheel is smoothed, and then the smoothed image is subtracted from the original image, and the resulting difference image is binarized; the binary image resulting from the low threshold is taken as the marker image, and that from the high threshold is taken as mask image, and mathematical morphology reconstruction is performed on the two images, with the resulting image being the preliminary result of the wheel defect segmentation. Finally, with area and diameter parameters as the conditions, the preliminary segmentation result is analyzed, and the defect regions satisfying the conditions are taken as the ultimate result of the whole solution. Experiments proved the feasibility of the above solution, which is found capable of extracting different types of wheel defects satisfactorily.

Keywords: aluminum alloy wheel; X-ray; nondestructive testing; defect detection; adaptive threshold; morphological reconstruction

1. Introduction

The aluminum alloy wheel is a main component and also a major load-bearing component of the car and its quality has an important effect on the overall performance of the car. At present, the mainstream production process of automobile hubs is low-pressure casting, in which the molten alloy liquid is poured into the ready mold. In the molding and cooling process, internal defects, like gas holes or shrinkage cavities or shrinkage porosity, occur if air does not escape in time or if the alloy liquid is not adequately replenished [1]. So, X-ray inspection equipment is required for flaw detection. A typical wheel X-ray inspection set includes a radiation source, a detector, a computer, a manipulator, and a protective lead-clad room. The radiation penetrates a certain part of the wheel and is picked up by the detector and then transmitted to the computer, where a corresponding X-ray image is generated. The manipulator moves the wheel so that every part of the wheel is exposed to the X-ray. The lead room is used to shield the X-ray radiation and protect the operators [2]. The X-ray image in the computer carries the internal defect information of the wheel. An important part of the X-ray inspection of the hub is to process and analyze the X-ray image.

Over the past two decades, a lot of literature appeared that dealt with defect segmentation and identification in wheel X-ray images. D. Merry and D. Filbert proposed a wheel defect detection technique based on video tracking [3]. It first used an edge detection operator to process the wheel image to get the preliminary detection result, and then, relying on area and mean gray, two quantities, to eliminate some pseudo-defects. Matching between and tracking of sequential images was performed on the remaining defects to produce the final detection result. In reference [4], X. Li et al. compared the second-order derivative and morphology operation, the row-by-row adaptive thresholding, and the two-dimensional (2-D) wavelet transform method. It was shown that only the 2-D wavelet transform was able to satisfactorily detect cracks, shrinkage cavities, and foreign inclusions, three defects mentioned in the paper [4]. Y. Tang et al. proposed a maximum fuzzy exponential entropy criterion based on bound histogram (MFEEC-BH) for extracting defects from wheel X-ray images, which made full use of the advantages of fuzzy set theory and bound histogram and was capable of fast and accurate separation of wheel defects from the background [5]. A. Osman et al., drawing on the Dempster–Shafer data fusion theory, used the measurement characteristics of the examined wheel as the information source to design an automatic defect-type recognition function by use of a manually classified database, which has improved the reliability of the wheel examination [6]. M. Carrasco, based on the fact that a physical hub defect is definable by geometries and is tracked in multiple images in an image sequence, designed a method that automatically detected the images of various parts of the hub from multiple viewpoints, which was able to detect true defects and pick out most pseudo-defects [7]. X. Zhao divided the detection of hub defects into two categories: direct detection and detection by model, and proposed to train the grayscale arranging pairs (GAP) feature on a series of images and then used the trained model to complete defect segmentation [8]. The Randomly Distributed Triangle (RDT) feature was extracted from the segmented defects, and a classifier called Sparse Representation-based Classification (SRC) was designed to classify the defects. This method enabled automatic recognition of weak and minor defects under low contrast and nonuniform illumination conditions [9]. Over recent years, with the advance of computer vision technology, some new methods have found their way into automatic detection of wheel hub defects. An example in point is the convolutional neural network technology used in deep learning [10].

Threshold segmentation, a classical image segmentation method, has the advantages of low computational cost and fast speed, and is widely used in the field of industrial X-ray examination [4,5,11–14]. A wheel is complex in geometry, the gray scale of its X-ray image varies widely, and a defect only accounts for a small part of the whole image, therefore accurately extracting defects by the traditional threshold segmentation is quite difficult. Considering the fact that a hub defect has a brighter gray than the background in the neighborhood, this paper makes use of a technique called 'adaptive threshold segmentation' to carry out defect segmentation by specifying the margin by which the target object is brighter than the background. To avoid the interference of noise and hub geometry on defect extraction, the segmentation result is processed by reconstruction operation, a technique in mathematical morphology. After processing, the area and diameter in all regions are counted, and proper area and diameter value ranges are determined having regard to the physical facts of the defects. Thus, accurate defect segmentation can be achieved.

2. Theoretical Background and Proposed Method

2.1. Adaptive Threshold Segmentation

The adaptive threshold segmentation algorithm is different from its traditional counterparts (e.g., OTSU algorithm [15]) which apply the same threshold to the entire image for segmentation. In contrast, the proposed algorithm applies a smoothing operator to the image, and then finds the difference between the original image and the smoothed one. Later, a fixed threshold is applied to the difference image, that is, how much the target object is brighter than the background, to achieve binarization. Suppose that $f(x,y)$ represents the original hub X-ray image to be segmented, $g(x,y)$ represents the image resulting from smoothing, and $B(x,y)$ represents the final segmentation result. Let T be the specified fixed threshold. Then, the adaptive threshold segmentation algorithm can be expressed as:

$$B(x,y) = [f(x,y) - g(x,y)] > T \tag{1}$$

Equation (1) has an equivalent expression as follows:

$$B(x,y) = f(x,y) > [g(x,y) + T] \tag{2}$$

For the original image $f(x,y)$, the actual threshold of each pixel (x,y) is the sum of the background gray value g, obtained by applying the smoothing operator to that point, and the specified threshold T, and so it varies with the background gray value g at that point. The gray value g on each pixel of the smoothed image $g(x,y)$ is jointly determined by the gray level of the corresponding pixels of the original image $f(x,y)$ and the peripheral pixels. Assuming that the smoothing operator is $h(x,y)$, then $g(x, y)$ is obtained from the following formula:

$$g(x,y) = f(x,y) * h(x,y) \tag{3}$$

The symbol "*" indicates a convolution operation in the digital signal processing. The smoothing operator $h(x, y)$ appears in the form of a matrix, usually being the mean smoothing operators and Gaussian smoothing operators. Taking the mean smoothing operator as an example, the $h(x, y)$ expression when the size r is 3 shall be

$$h(x,y) = \frac{1}{3 \times 3} \begin{bmatrix} 1 & 1 & 1 \\ 1 & 1 & 1 \\ 1 & 1 & 1 \end{bmatrix} \tag{4}$$

Using the operator in Formula (4), the original image $f(x, y)$ is operated according to Formula (3), and the gray value g of each point in the image $g(x, y)$ that has been obtained is the average gray scale of a total of 9 pixels in a 3×3 square area that takes that point as the center. Figure 1a shows a part of the gray value of the X-ray image of the hub, and the image $g(x, y)$ obtained through 3×3 mean smoothing is shown in Figure 1b. Assuming that the fixed threshold T is set to 5, the actual threshold value of each point in Figure 1a when it is processed by binarization based on Formula (2) is the gray value corresponding to that point in Figure 1b plus 5, that is, Figure 1c. Finally, the binarization result of Figure 1a is equivalent to selecting a target object whose gray scale is larger than the average background by 5 in 3×3 local area, as shown in Figure 1d.

Figure 2a shows a part of a hub X-ray image. There is an obvious shrinkage cavity in this part, but it is of a small proportion of the image. It is also noted that the background gray of the image varies widely. The image is segmented with the OTSU algorithm and the threshold is found to be 162, so each pixel with a gray value greater than 162 is taken as a pixel from a target object. Those with a value below 162 are taken as one from the background. The segmentation result is shown in Figure 2b. As can be seen, the target object, or the shrinkage cavity, has not been accurately extracted. The image obtained by smoothing Figure 2a with a 25×25 mean filter is shown in Figure 2c, in which the gray value of each pixel is the mean of a 25×25 square centered on this pixel in Figure 2a. From Equations (1) or (2), with the threshold T taken as 2, the resulting segmentation is as shown in Figure 2d. It is clear that the adaptive threshold segmentation produces a much better result than the OTSU algorithm, because the defect in the image has been segmented with accuracy. The reason is that the adaptive threshold segmentation algorithm takes advantage of the fact that a target object is brighter among its local background. In regions without a target object, the gray scale changes in a graded manner, with the result that the original gray scale differs not much from the smoothed one (less than or equal to T), and is then regarded as the background. In contrast, a region with a target object changes drastically in gray value, with the result that the original gray differs significantly from the smoothed one (greater than *T*), and is then regarded as an object. The adaptive threshold segmentation algorithm focuses on local gray variation and is therefore more robust than fixed threshold segmentation algorithms when the target objects are smaller and the background gray is more complex, but undesirably, the noise points and the edges of light regions or dark regions in the image, which change drastically too in gray, are segmented out too. As shown in Figure 2d, some portions not related to defects are segmented out.

45	60	98	123	133	137	139	133
46	79	82	93	105	120	112	124
47	56	83	95	108	106	101	103
47	82	91	99	103	101	100	96
50	93	95	92	102	108	99	85
49	84	81	73	97	105	82	78
50	65	68	68	81	85	74	73
50	41	38	37	42	46	49	64

(a)

52	68	91	110	123	128	130	130
52	66	85	102	113	118	119	119
55	68	84	95	103	106	107	107
58	72	87	96	102	103	100	96
61	75	88	93	98	100	95	89
60	71	80	84	90	93	88	81
54	58	62	65	70	73	73	71
50	49	48	50	54	57	61	64

(b)

57	73	96	115	128	133	135	135
57	71	90	107	118	123	124	124
60	73	89	100	108	111	112	112
63	77	92	101	107	108	105	101
66	80	93	98	103	105	100	94
65	76	85	89	95	98	93	86
59	63	67	70	75	78	78	76
55	54	53	55	59	62	66	69

(c)

0	0	1	1	1	1	1	0
0	1	0	0	0	0	0	0
0	0	0	0	0	0	0	0
0	1	0	0	0	0	0	0
0	1	1	0	0	1	0	0
0	1	0	0	1	1	0	0
0	1	1	0	1	1	0	0
0	0	0	0	0	0	0	0

(d)

Figure 1. Example of adaptive threshold: (**a**) original image; (**b**) smoothed image; the light green pixels indicate the source neighborhood for the light green destination pixel; (**c**) true thresholds of original image when *T* = 5; (**d**) result of adaptive threshold.

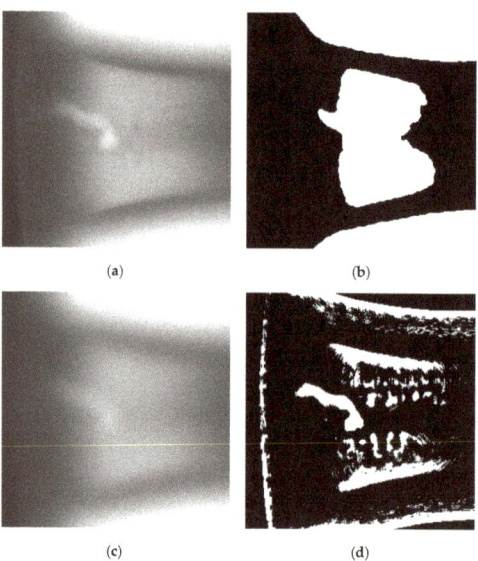

Figure 2. Segmentation results using adaptive threshold and OTSU methods: (**a**) original image with defect; (**b**) result of OTSU method; (**c**) smoothing image by a 25 × 25 mean filter; (**d**) result of adaptive threshold method.

The above analysis shows that the smoothing factor size and the threshold value are two determining factors in the adaptive threshold segmentation algorithm. Suitable operators for image smoothing are mean filtering, Gaussian filtering, or median filtering operators, the size of which is the size of the local area. This size determines the size of the objects that can be segmented. Too small a filter size is unable to give an accurate estimate of the local background brightness at the center of the object, resulting in segmentation failure. The larger the filter size, the better the filtered result will represent the local background, and the more likely the object is accurately segmented out. But too large a filter size will result in higher computational load, and adjacent objects, too, may have an undesirable effect on the filtering results. Experience suggests that when the filter is about the size of the object to be recognized, an accurate estimation of the background gray level of the defect and an accurate segmentation of the defect can be obtained at once. The value of the threshold T varies with the object to be segmented. A larger threshold suppresses the noise better, but may lead to the loss of the edge pixels of the target object, resulting in incomplete segmentation. A smaller threshold ensures that the target object is completely segmented, but noise and light and dark edges may exert some influence. For Figure 2a, the segmentation results of a smaller and a larger filter size but with the T value maintained at 2 are shown in in Figure 3a,b, respectively, while the segmentation results of a smaller T value and a larger T value but with the filter size maintained unchanged are shown in in Figure 3c,d, respectively. Figure 3a shows the segmentation result when the filter for smoothing is set at 9×9, the smaller size. Too small a filter window leads to local background estimation inaccuracy, and compared with Figure 2d, the defect is not a whole one but consists of discrete pieces. Figure 3b shows the segmentation result when the filter size is increased to 51×51. Although the defect is segmented out as a whole compared with Figure 2d, too large a filtering window makes the hub geometry interfere with the defect area, such that the two come together, resulting in segmentation failure. Figure 3c shows the segmentation result when the threshold T is set to 0, and the noise interference and the effect of the light and dark edges are significantly stronger than in Figure 2d. Figure 3d shows the segmentation result when the threshold T becomes 20, and it can be seen that both noise and edge interference disappear, but the defect is just partially segmented out.

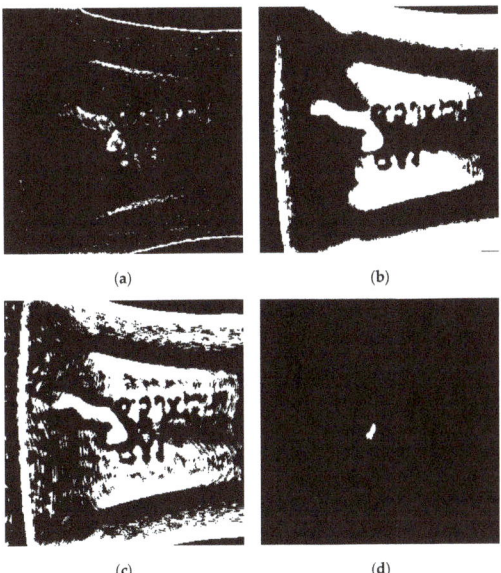

Figure 3. Results of adaptive threshold with different parameters: (**a**) result of the smaller size filter; (**b**) result of the bigger size filter; (**c**) result of the smaller threshold; (**d**) result of the bigger threshold.

It becomes clear now that too large or too small a smoothing window size or threshold value affects the final segmentation result. The two parameters, smoothing window size and threshold value, have to be determined intelligently to bring about perfect segmentation results. However, in practical applications, it is hard to achieve this goal, especially in defect detection of wheel X-ray images.

2.2. Morphological Reconstruction

Mathematical morphology originated from the geometric study of the permeability of porous media by French scholars in the 1960s. It was initially confined to the geometrical analysis of binary images, but slowly expanded to the field of grayscale and color images. Mathematically, it is based on set theory, integral geometry, and mesh algebra. It has gradually developed into a powerful image analysis technology, widely used in the field of industrial nondestructive testing [16,17]. Mathematical morphology detects images through a small set called structuring element, and its basic operations include dilation and erosion. From this as the basis, other transformations are made through combinations. Reconstruction of mathematical morphology involves two images, one called a mask image and the other called a marker image, with the latter being smaller than or equal to its corresponding mask image. Reconstruction transformation is an iterative process in which the marker image is used to reconstruct the mask image. The operation process begins with dilating the marker image using 3 × 3 all one-square structuring elements, and the dilation result is compared to the mask image point by point and the ones with the lower value are taken as the intermediate result. This intermediate result then replaces the marker image to start another round of dilation and point-by-point comparison with the mask image, and similarly, the ones with the lower value are taken as the intermediate result. This iteration continues until the intermediate result changes no more, and this intermediate result is taken as the final reconstruction result. Let $m(x,y)$ represent the marker image, $f(x,y)$ represent the mask image, and R represent the reconstruction process. Then, the morphological reconstruction operation can be expressed as:

$$f_R(x,y) = R_{f(x,y)}[m(x,y)] \tag{5}$$

where $f_R(x,y)$ is the result of the reconstruction operation. Reconstruction operation attempts to restore the mask image $f(x,y)$ using the marker image $m(x,y)$, and the light regions on $m(x,y)$ that have completely disappeared will not be restored in $f_R(x,y)$, but light regions partially shown in $m(x,y)$ are fully recovered in $f_R(x,y)$. The result of reconstruction operation, with the image in Figure 3d as the marker image $m(x,y)$ and the image in Figure 2d as the mask image $f(x,y)$, is as shown in Figure 4. It can be seen that the defect region is completely recovered, and noise and edge interference is removed too, producing a perfect defect segmentation.

Figure 4. Result of morphology reconstruction.

If the high-threshold segmentation image is used to reconstruct the low-threshold segmentation image in the adaptive threshold segmentation algorithm, then the reconstruction result will completely recover the low-threshold segmentation regions tagged by the high-threshold segmentation, completely removing the low-threshold segmentation regions not tagged by the high-threshold segmentation. In other words, morphological reconstruction operation combines the advantages of high-threshold segmentation (low interference) and low-threshold segmentation (complete defects), and thus qualifies as a useful supplement to adaptive threshold segmentation algorithm and minimizes the difficulty of parameter setting for segmentation.

2.3. Procedures of the Proposed Method

Drawing on adaptive threshold segmentation algorithm and morphological reconstruction operation, the proposed hub defect segmentation solution consists of the following steps:

1. Choose a smoothing operator of a suitable size to smooth the wheel X-ray image to obtain a smoothed image.
2. The smoothed image is subtracted from the original image to obtain a difference image.
3. Choose a smaller threshold value for the difference image to perform binarization to obtain the first-time segmentation result, and the result is used as a mask image for morphological reconstruction.
4. Choose a larger threshold value for the difference image to perform binarization to obtain the segmentation result, and the result is used as the marker image for morphological reconstruction.
5. Perform morphological reconstruction using the marker image and the mask image to obtain the preliminary defect segmentation result.
6. Perform preliminary analysis of the defect segmentation result having regard to the physical facts of the wheel defect, and this produces the final defect segmentation result.

Throughout the segmentation process, the parameters involved include the size of the smoothing operator and the threshold value for binarization, the determination of which has been discussed in Section 2.1. In step 3, the small threshold binarization produces whole defect regions, but pseudo-defects such as noise and structural interference are inevitable. In step 4, the large threshold binarization only produces partial region of defects, but most pseudo-defects are removed. The morphological reconstruction operation in step 5 restores complete defect regions obtained in step 3 but removes the pseudo-defects not tagged in step 4, to produce the preliminary segmentation result. For many wheel images, the geometry of the hub itself creates sharp edges in the X-ray image, and such edges cannot be removed by steps 3, 4, and 5, but their geometric features, such as area and diameter, are significantly larger than common defects. Step 6 includes further analysis of the preliminary result obtained in step 5 with a view to eliminating such pseudo-defects by studying the geometric features of the defect regions, and the product of this step is the final segmentation result.

3. Experiment Results

The data for this experiment came from a wheel defect on-line detection system which includes an X-ray detector as an image intensifier plus a CCD camera. The resolution of the camera is 768 × 576, and the power of the ray source is set to 170 kV and 2 mA. Figure 5a shows an X-ray image of a wheel acquired by this system, and Figure 2a shows a part of the image. Given the state-of-the-art wheel hub production, hub defects are generally small. A smoothing operator of size r about 5% of the long side of the acquired image will satisfy the detection purpose and, in this system, the size is about 39. The ray source by its voltage and current determines the gradation change of the X-ray image. For the combination of 170 kV and 2 mA, the small threshold T_s in step 3 of the segmentation process should be set between 0 and 10, and the large threshold T_b in step 4 should be between 10 and 20. The area s of the hub defect is defined as the number of pixels in the segmented region, and the diameter d is the maximum distance between any two points on the edge of the segmented region. According to the physical fact that a hub defect is typically small, the upper limit M_s of the defect area is generally smaller than 10% of the total number of pixels in the image, and the upper limit M_d of the diameter is generally less than 20% of the shorter side of the entire image. In the acquisition system in question, M_s is 4424 and M_d is 115. In step 6 of the division process, a region having an area smaller than M_s and having a diameter smaller than M_d is retained as the final defect region.

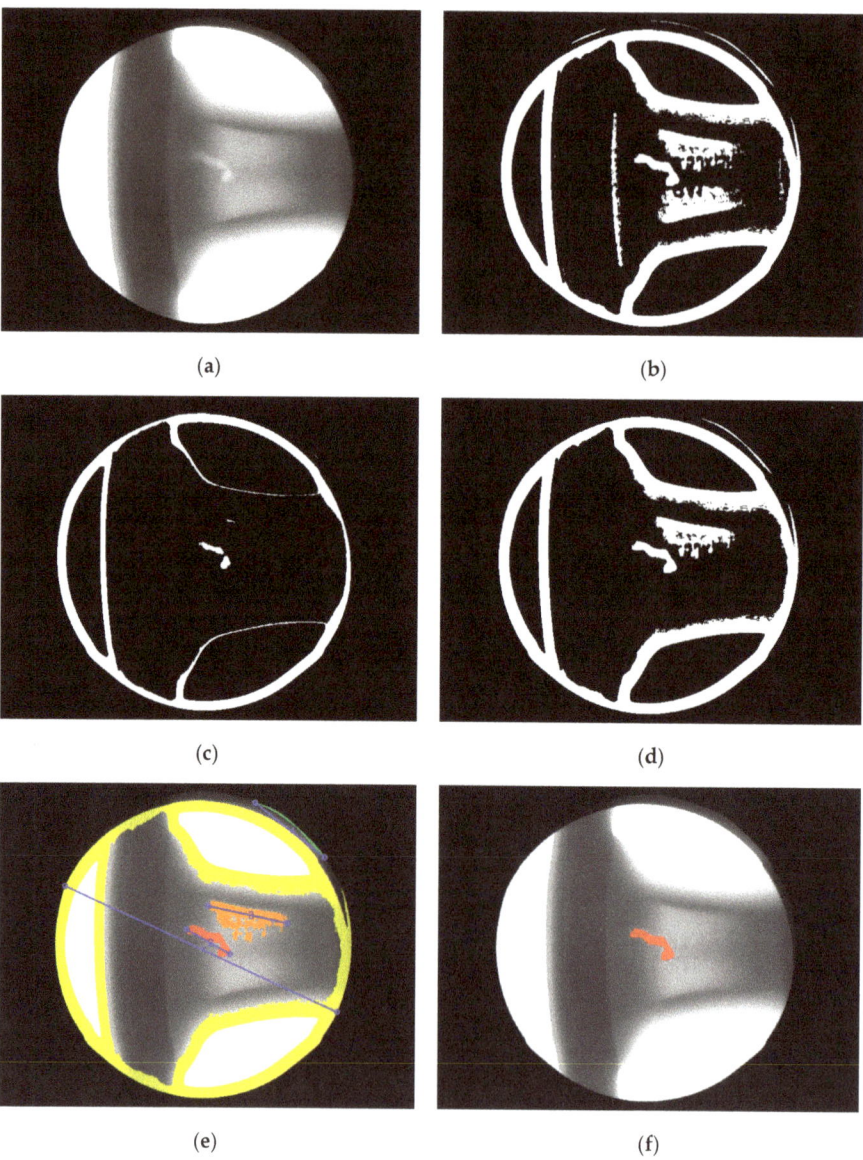

Figure 5. Segmentation result of proposed method: (**a**) original image; (**b**) result of $T_s = 1$; (**c**) result of $T_b = 12$; (**d**) result of reconstruction; (**e**) pseudo-color image of preliminary defects; (**f**) final result of proposed method.

Figure 5a is smoothed using a mean smoothing operator with an r of 39. The segmentation result with the small threshold T_s set to 1 is shown in Figure 5b, and the segmentation result with the large threshold T_b set to 12 is shown in Figure 5c. The morphological reconstruction result is shown in Figure 5d. The reconstruction result includes the sought defect region along with pretty much interference. So, the result is to be analyzed having regard to the area and diameter, as described in step 6 of the segmentation process. Figure 5d is superimposed, in the format of a pseudo-color image, onto the original image Figure 5a, with its diameter displayed and each region numbered, and Figure 5e shows the result. Each defect region is studied for its area and diameter, and the data are shown in Table 1. As can be seen from Table 1, region 2, which meets the requirement that its area shall be smaller than M_s and its diameter shall be smaller than M_d, is the true defect region. Performing step 6 of the segmentation process to get the defect region, which is marked on the original image, as shown in Figure 5f, it can be seen that the defect is accurately segmented out while the interference is altogether removed.

Table 1. Parameters of preliminary defects.

No.	Area	Diameter	Less than M_s?	Less than M_d?	A Real Defect?
1	57,867	546.7	N	N	N
2	1563	90.8	Y	Y	Y
3	4711	149.1	N	N	N
4	462	159.2	Y	N	N

To verify the performance of the algorithm, testing was conducted on different hub images. Figure 6a is another hub X-ray image with defects, taken from the same acquisition system. Figure 6b shows the segmentation result obtained from the same parameters, that is, the size r of the mean smoothing operator is 39, the small threshold T_s is 1, the large threshold T_b is set to 12, the defect area is less than 4424, and its diameter is less than 115. It can be seen that the more obvious large defect is extracted with accuracy while the small one, which is less obvious and of a poor contrast, is also extracted.

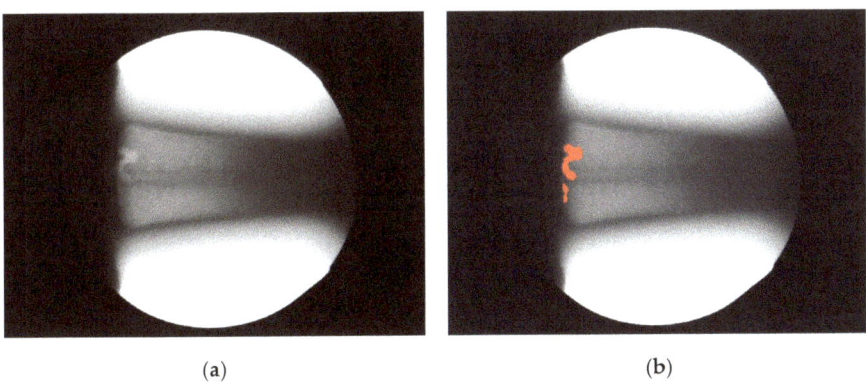

(a)   (b)

Figure 6. Another result of proposed method: (a) original image; (b) result of proposed method.

Figure 7a shows still another hub X-ray image with small defects. The defects were segmented using the same parameters as in Figure 6a. The initial segmentation result, produced by step 3, is shown in Figure 7b. As can be seen, the defect regions are not distinguishable from those created by the hub geometry, so subsequent steps are unable to extract the defects correctly. The smoothing operator was resized to have an r of 19, and the small threshold T_s was raised to 5, to have produced the initial segmentation result as shown in Figure 7c. As can be seen, the defects are extracted, distinguishable from interference. Continuing from Figure 7c, with the large threshold T_b, the area M_s, and the diameter M_d unchanged, the remaining steps of the segmentation process were performed, to have produced the final defect segmentation result as shown in Figure 7d, in which the defects are marked on the original image. The defects are accurately segmented out.

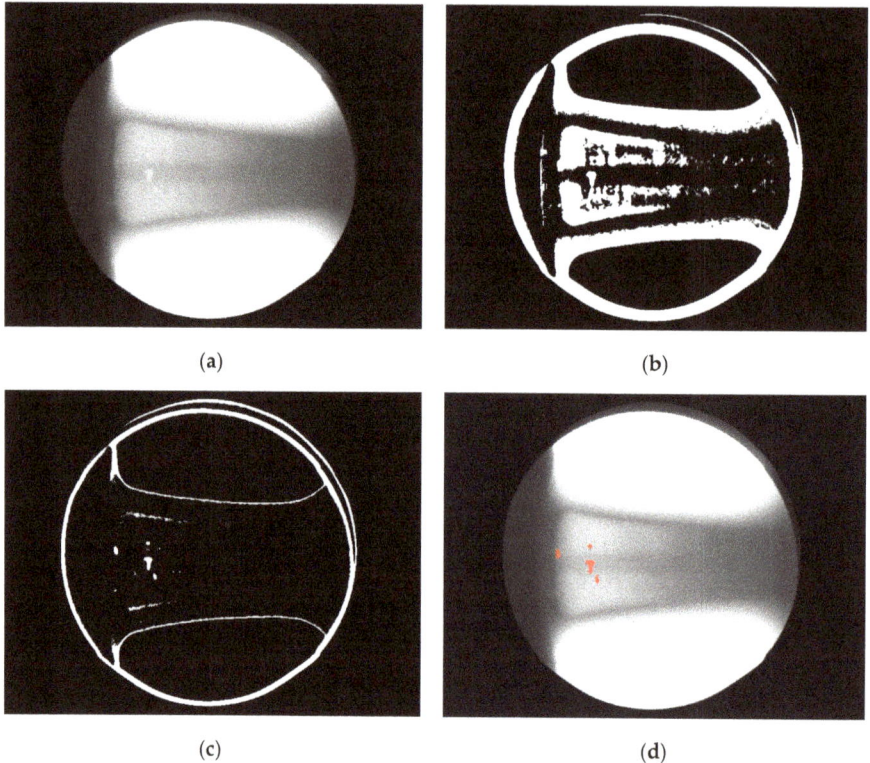

Figure 7. Another result of proposed method: (a) original image; (b) segmentation result of $r = 39$ and $T_s = 1$; (c) segmentation result of $r = 19$ and $T_s = 5$; (d) final result of proposed method.

4. Discussion

This paper proposes a technical solution that combines adaptive threshold segmentation algorithm and morphological reconstruction operation to extract the defects on wheel X-ray images. The innovation of the algorithm for dynamic threshold segmentation is that in case of defect segmentation, the algorithm focuses on the grayscale variation of the local area, and the size of local area and the grayscale variation can be directly determined by setting the parameters r and T, which is very suitable for the extraction of the hub defect. The morphological reconstruction operation restores the mask image by specifying the marking image. The operation feature is that the area existing in the marking image can be completely restored in the mask image, and the area not in the marking image will completely disappear in the mask image. Combining the algorithm for dynamic threshold segmentation with the morphological reconstruction operation is the max novelty in this paper. Taking the high-threshold segmentation result in the dynamic threshold segmentation algorithm as the marking image, and the low-threshold segmentation result as the mask image, the defect area marked by the high-threshold segmentation result after the reconstruction operation can be completely restored, and the interference area generated by the low-threshold segmentation result can be completely removed, and the accurate extraction of the hub defect is realized.

Parameters shall be set to match different types and sizes of defects. The choice of the smoothing operator size r and the small threshold T_s is critical, for it determines what defects will be finally segmented out. Parameters shall be such that the defects are all segmented out and separated from the interference at once. Improper parameters, examples of which are shown in Figures 3 and 7, may lead to failure in defect extraction. Large threshold T_b, area M_s, and diameter M_d are designed to remove pseudo-defects. Parameter T_b shall be such that maximum interference is discarded without ignoring the defects. Parameter M_s and M_d act to remove the interference that is generated by the hub geometry and is not removable by thresholding. The experiment results show that the proposed method is capable of accurate segmentation of wheel hub defects and in practical applications, it is important to make proper parameter settings.

Author Contributions: Conceptualization, J.Z. and M.W.; Methodology, J.Z.; Software, Z.G. and T.J.; Validation, J.Z., Z.G., and T.J.; Formal Analysis, T.J.; Investigation, Z.G. and T.J.; Resources, M.W.; Data Curation, Z.G. and T.J.; Writing—Original Draft Preparation, J.Z.; Writing—Review & Editing, J.Z. and M.W.; Visualization, J.Z.; Supervision, M.W.; Project Administration, J.Z. and M.W.; Funding Acquisition, M.W.

Funding: This research was funded by National Special Project for the Development of Major Scientific Instruments and Equipment of China (No. 2013YQ240803) and Scientific and Technological Innovation Programs of Higher Education Institutions in Shanxi Province (2013163).

Conflicts of Interest: The authors declare no conflict of interest.

References

1. Zhang, B.; Cockcroft, S.L.; Maijer, D.M.; Zhu, J.D.; Phillion, A.B. Casting defects in low-pressure die-cast aluminum alloy wheels. *JOM* **2005**, *57*, 36–43. [CrossRef]
2. Mery, D.; Jaeger, T.; Filbert, D. A review of methods for automated recognition of casting defects. *Insight* **2002**, *44*, 428–436.
3. Mery, D.; Filbert, D. Automated flaw detection in aluminum castings based on the tracking of potential defects in a radioscopic image sequence. *IEEE Trans. Robot. Autom.* **2002**, *18*, 890–901. [CrossRef]
4. Li, X.; Tso, S.K.; Guan, X.; Huang, Q. Improving Automatic Detection of Defects in Castings by Applying Wavelet Technique. *IEEE Trans. Ind. Electron.* **2006**, *53*, 1927–1934. [CrossRef]
5. Tang, Y.; Zhang, X.; Li, X.; Guan, X. Application of a new image segmentation method to detection of defects in castings. *Int. J. Adv. Manuf. Technol.* **2009**, *43*, 431–439. [CrossRef]
6. Osman, A.; Kaftandjian, V.; Hassler, U. Improvement of x-ray castings inspection reliability by using Dempster–Shafer data fusion theory. *Pattern Recognit. Lett.* **2011**, *32*, 168–180. [CrossRef]
7. Carrasco, M.; Mery, D. Automatic multiple view inspection using geometrical tracking and feature analysis in aluminum wheels. *Mach. Vis. Appl.* **2011**, *22*, 157–170. [CrossRef]
8. Zhao, X.; He, Z.; Zhang, S. Defect detection of castings in radiography images using a robust statistical feature. *J. Opt. Soc. Am. A* **2014**, *31*, 196–205. [CrossRef] [PubMed]
9. Zhao, X.; He, Z.; Zhang, S.; Liang, D. A sparse-representation-based robust inspection system for hidden defects classification in casting components. *Neurocomputing* **2015**, *153*, 1–10. [CrossRef]
10. Mery, D.; Arteta, C. Automatic Defect Recognition in X-ray Testing using Computer Vision. In Proceedings of the 2017 IEEE Winter Conference on Applications of Computer Vision (WACV), Santa Rosa, CA, USA, 24–31 March 2017.
11. Saravanan, T.; Bagavathiappan, S.; Philip, J.; Jayakumar, T.; Rai, B. Segmentation of defects from radiography images by the histogram concavity threshold method. *Insight* **2007**, *49*, 578–584. [CrossRef]
12. Wang, Y.; Sun, Y.; Lv, P.; Wang, H. Detection of line weld defects based on multiple thresholds and support vector machine. *NDT E Int.* **2008**, *41*, 517–524. [CrossRef]
13. Yuan, X.; Wu, L.; Peng, Q. An improved Otsu method using the weighted object variance for defect detection. *Appl. Surf. Sci.* **2015**, *349*, 472–484. [CrossRef]
14. Malarvel, M.; Sethumadhavan, G.; Bhagi, P.C.R.; Kar, S.; Thangavel, S. An improved version of Otsu's method for segmentation of weld defects on X-radiography images. *Optik* **2017**, *142*, 109–118. [CrossRef]
15. Otsu, N. A threshold selection method from gray-level histograms. *IEEE Trans. Syst. Man Cybern.* **1979**, *9*, 62–66. [CrossRef]

16. Alaknanda, R.; Anand, S.; Kumar, P. Flaw detection in radio-graphic weld images using morphological approach. *NDT E Int.* **2006**, *39*, 29–33. [CrossRef]
17. Alaknanda, R.; Anand, S.; Kumar, P. Flaw detection in radio-graphic weldment images using morphological watershed segmentation technique. *NDT E Int.* **2009**, *42*, 2–8. [CrossRef]

© 2018 by the authors. Licensee MDPI, Basel, Switzerland. This article is an open access article distributed under the terms and conditions of the Creative Commons Attribution (CC BY) license (http://creativecommons.org/licenses/by/4.0/).

Article

Non-Contact Ultrasonic Inspection of Impact Damage in Composite Laminates by Visualization of Lamb wave Propagation

Nobuyuki Toyama [1,*], Jiaxing Ye [1], Wataru Kokuyama [1] and Shigeki Yashiro [2]

[1] National Metrology Institute of Japan, National Institute of Advanced Industrial Science and Technology (AIST), 1-1-1 Umezono, Tsukuba, Ibaraki 305-8568, Japan; jiaxing.you@aist.go.jp (J.Y.); wataru.kokuyama@aist.go.jp (W.K.)
[2] Department of Aeronautics and Astronautics, Kyushu University, 744 Motooka, Nishi-ku, Fukuoka 819-0395, Japan; yashiro@aero.kyushu-u.ac.jp
* Correspondence: toyama-n@aist.go.jp; Tel.: +81-29-861-3025

Received: 27 November 2018; Accepted: 21 December 2018; Published: 24 December 2018

Abstract: This study demonstrates a rapid non-contact ultrasonic inspection technique by visualization of Lamb wave propagation for detecting impact damage in carbon fiber reinforced polymer (CFRP) laminates. We have developed an optimized laser ultrasonic imaging system, which consists of a rapid pulsed laser scanning unit for ultrasonic generation and a laser Doppler vibrometer (LDV) unit for ultrasonic reception. CFRP laminates were subjected to low-velocity impact to introduce barely visible impact damage. In order to improve the signal-to-noise ratio of the detected ultrasonic signal, retroreflective tape and a signal averaging process were used. We thus successfully visualized the propagation of the pulsed Lamb A_0 mode in the CFRP laminates without contact. Interactions between the Lamb waves and impact damage were clearly observed and the damage was easily detected through the change in wave propagation. Furthermore, we demonstrated that the damage could be rapidly detected without signal averaging. This method has significant advantages in detecting damage compared to the conventional method using a contact resonant ultrasonic transducer due to the absence of the ringing phenomenon when using the LDV.

Keywords: non-destructive inspection; laser ultrasonic imaging; Lamb wave; delamination; composite laminate

1. Introduction

Carbon fiber reinforced polymer (CFRP) laminates are increasingly being applied to structural components in aircrafts and automobiles to improve fuel efficiencies, due to its lightweight, superior strength and stiffness. Composite structures in these safety-critical applications must be inspected to ensure safety and reliability and to prevent catastrophic failure. Among the various types of damage, internal damage from low-velocity impact is the most common type found in composite structures. This damage is easily induced from things as simple as tools being dropped during maintenance. Damage presents in the form of matrix cracks, delamination, and fiber breakage. Moreover, this damage is barely visible to the naked eye on the structure's surface, which is explained by the term, "barely visible impact damage (BVID)". Delamination in particular, must be detected during inspection processes, as it causes a significant loss of compressive strength. Current inspection practices employ non-destructive testing (NDT) techniques such as X-ray or ultrasonic C-scan to identify delamination. However, these techniques are very time-consuming and expensive for inspecting large structures. Therefore, a new, non-contact NDT technique to detect damage quickly, reliably, and automatically is required by industry. Should any damage be found, the conventional method can be applied to evaluate the damage in detail.

Ultrasonic waves propagate as Lamb waves in thin plate-like structures such as aircraft skins and automobiles bodies. They have significant potential for large-area, non-destructive inspection because they have a long propagation distance and allow the whole volume of the material between the transmitting and receiving transducers to be inspected. Therefore, Lamb wave inspection has been extensively applied for the detection of delamination in CFRP laminates in the literatures [1–8]. Recently, phased array ultrasonic techniques have also been developed for large-area inspection [9,10]. However, interpretation of the detected Lamb waves is challenging due to their dispersive nature, as well as the presence of multiple modes and scattered waves from the edge of the components.

On the other hand, by visualizing the ultrasonic waves propagating in an actual structure, the appearance of the additional scattered waves can be used to directly observe damage—without having to interpret the complicated measured waveforms. Ultrasonic wave propagation visualization is thus very effective for reliable damage inspection. We have previously developed a technique for the visualization of ultrasonic wave propagation in general solid media [11], which uses a pulsed laser that scans an object for ultrasonic wave generation and a fixed contact receiver to provide a movie (or series of snapshots) of the propagating waves. Although this is not a "fully" non-contact technique, it operates excellently, enables quick inspection of objects of arbitrary shapes. We have applied it to the non-destructive inspection of various structural components and demonstrated its usefulness [12–14].

In recent times, the development of ultrasonic wavefield imaging techniques for detecting delamination in composite laminates has attracted notable attention. Measurement systems utilizing a combination of fixed and/or scanning sources and receivers have been proposed to obtain the Lamb wave propagation images [15–22]. However, limited research has been undertaken on complete non-contact ultrasonic wavefield imaging techniques. Park et al. [20] adopted a Nd: YAG pulse laser for ultrasonic wave generation and a laser Doppler vibrometer (LDV) for reception to obtain the wavefield of the Lamb wave in composite structures, and detected delamination and disbonding. However, further studies are still needed to improve, especially in relation to inspection time, inspection area and image quality.

In this study, we demonstrate a rapid non-contact ultrasonic inspection technique by visualization of Lamb wave propagation for detecting BVID in CFRP laminates. This optimized laser ultrasonic wavefield imaging system utilizes rapid pulsed laser scanning and LDV units to clearly visualize damage in impacted CFRP laminates. Measurement techniques are developed to improve the signal-to-noise ratio of the detected ultrasonic signals. Furthermore, we compare the proposed method to a conventional contact piezoelectric transducer method. Through this study, we demonstrate the efficiency and feasibility of the proposed technique for the non-contact inspection of composite structures.

2. Experimental Procedure

2.1. Specimens

The materials used were CFRP (TR380-G250SM, Mitsubishi Chemical, Tokyo, Japan) cross-ply and quasi-isotropic laminates with stacking sequences of $[0/90]_{2S}$ and $[0/45/90/-45]_S$. The specimens had dimensions of $160 \times 160 \times 1$ mm and were subjected to low-velocity impact with an energy of 6 J using a vertical drop-weight impact system (CREAST 9310, Instron, Norwood, MA, USA) to induce the BVID. A hemispherical impactor with a diameter of 20 mm was used. The impacted specimens were inspected using a water-immersion ultrasonic C-scan system (TT-UTCS01, Tsukuba Technology, Tsukuba, Japan). Square regions of 40×40 mm, which included the impacted positions, were scanned with an interval of 0.3 mm using a focused ultrasonic transducer with a resonant frequency of 10 MHz, diameter of 5 mm, and focal distance of 25 mm. Figure 1 shows the C-scan images for both specimens. Internal impact damages consisting of delamination of multiple interlayers were clearly detected, with dimensions of about 14 mm in the major axis and 9 mm in the minor axis. These BVIDs are the inspection targets of this study.

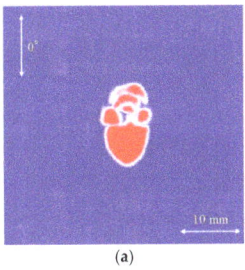
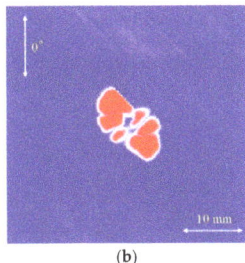

(a)   (b)

Figure 1. C-scan images depicting impact-induced delamination for carbon fiber reinforced polymer (CFRP) laminates subjected to impact loading with an energy of 6 J. (**a**) $[0/90]_{2S}$, (**b**) $[0/45/90/-45]_{S}$.

2.2. Non-Contact Laser Imaging System for Ultrasonic Wave Propagation

Figure 2 displays a photograph of the non-contact laser imaging system for visualizing ultrasonic wave propagation. This system consists of a rapid pulsed laser scanning unit for ultrasonic generation and a LDV unit for ultrasonic reception. Pulsed thermoelastic ultrasonic waves are generated by illuminating the specimen surface with a Q-switched Nd: YAG laser (Wedge-HB-1064-DB, Bright Solutions, Pavia, Italy) with a wavelength of 1064 nm, pulse width of 1.5 ns, maximum pulse energy of 2 mJ, and maximum repetition frequency of 2 kHz. The diameter of the laser beam is reduced by using a varifocal lens (APL-1050, Holochip, Hawthorne, CA, USA). The laser beam is scanned on the specimen surface using a computer-controlled galvanometer mirror (VM500+, Novanta, Bedford, MA, USA). A green laser beam with a wavelength of 532 nm is also illuminated for convenience, since the pulsed laser beam is invisible. The LDV system is used to receive the ultrasonic wave signals. It consists of a modular vibrometer (OFV-5000, Polytec, Waldbronn, Germany) and a sensor head (OFV-505-KA, Polytec, Waldbronn, Germany). A He-Ne continuous wave (CW) laser with a wavelength of 633 nm and energy of 2 mW is illuminated at a fixed position on the specimen surface, and the out-of-plane displacement at that position is measured based on the Doppler effect. The received signals are bandpass-filtered from 50 to 400 kHz using a variable-frequency filter (3628 Dual Channel Programmable Filter, NF, Yokohama, Japan) and stored in the computer through a high-speed digitizer (NI PCI-5124, National Instruments, Austin, TX, USA). A snapshot of the propagating waves at any given time is obtained by plotting the amplitude of each waveform at that time on a contour map. The snapshots can be continuously displayed in a time series to form a video of the waves propagating beneath the CW laser.

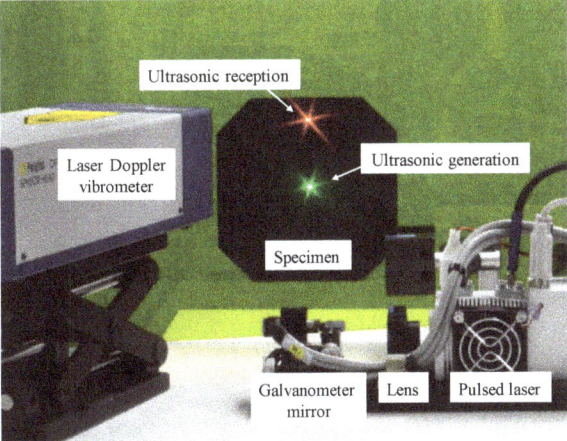

Figure 2. Photograph of the non-contact laser imaging system for visualizing ultrasonic wave propagation.

2.3. Non-Contact Ultrasonic Inspection

The specimens were vertically fixed at a position about 400 mm from the galvanometer mirror and LDV sensor head. As illustrated in Figure 3, laser scanning with an interval of 0.5 mm was performed in square regions of 80 × 80 mm. To improve the signal-to-noise ratio, each position was illuminated 30 times during the scanning and the 30 received signals were averaged. The pulse energy and scanning speed were set to about 0.6 mJ and 300 points/s, respectively, so that the laser illumination did not cause surface ablation. Due to the poor reflectivity of the CW laser, and the fact that the diameter of the laser beam at the specimen surface was about 50 µm, a 3 × 3 mm square of retroreflective tape (A-RET-T010, Polytec, Waldbronn, Germany) was attached to the specimen surface at the CW-laser illuminated position (40 mm from the impacted position) to improve the reflectivity. It should be noted that the scanning speed used in this study is estimated to be about 25 times faster than that in the previously reported technique [20,21].

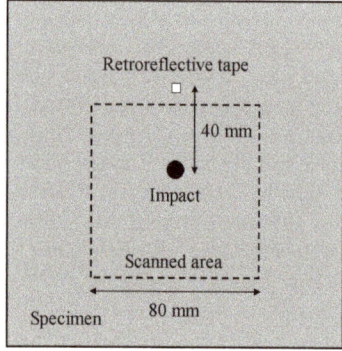

Figure 3. Schematic of the pulsed laser scanning area and the retroreflective tape position on the impacted specimen.

3. Results

3.1. Non-Contact Ultrasonic Damage Inspection

Figure 4 depicts the visual results of the ultrasonic wave propagation for both specimens. The ellipses shown in the top figures demonstrate the approximate location and shape of the damage in each specimen. The ultrasonic waves propagate as Lamb waves in these thin CFRP laminates, and it was confirmed that only the first symmetric (S_0) and first anti-symmetric (A_0) modes exist in the low frequency range between 50 and 400 kHz by using the dispersion curve analysis program Disperse [23]. The faster S_0 mode is almost invisible in the figure since the amplitude detected by the LDV was very low. On the other hand, the detected amplitude of the slower A_0 mode was high enough for its propagation behavior with a pulsed shape to be clearly visualized. In both specimens, when the A_0 mode reaches the damage, a phase delay is observed. Furthermore, following the phase delay, reflected waves from the damage are distinctly observed; the shape of the reflected wave strongly depends on that of the damage shown in Figure 1. It is very important that the detailed interactions between the Lamb wave and the impact-induced damage were clearly visualized, and that the damage was easily detected using the proposed non-contact laser imaging method for ultrasonic-wave propagation. These results demonstrate the drastic improvement in inspection time without losing the quality in the ultrasonic images [20,21].

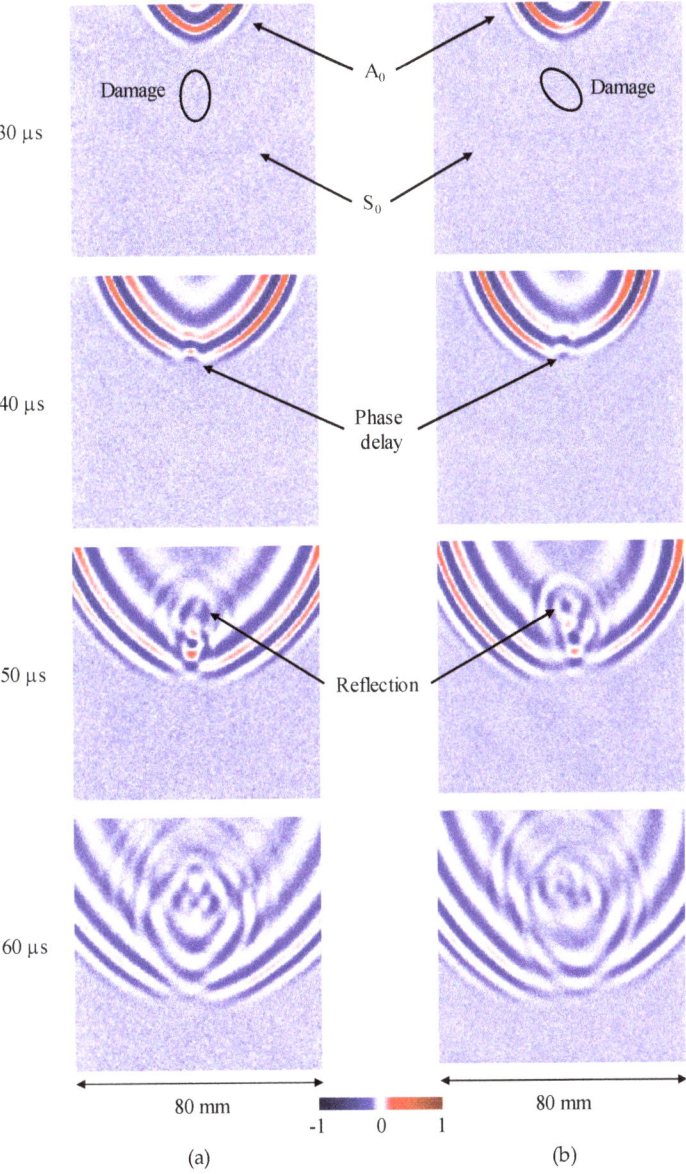

Figure 4. Lamb wave propagation in the impacted carbon fiber reinforced polymer (CFRP) laminates using the non-contact laser imaging system. Interactions between Lamb waves and impact-induced damages are clearly visible. (**a**) $[0/90]_{2S}$, (**b**) $[0/45/90/-45]_{S}$.

3.2. Rapid Ultrasonic Damage Detection

In order to achieve non-contact inspection by visualization of ultrasonic wave propagation, it is necessary to use the LDV to receive the low-energy ultrasonic waves induced by the pulsed laser. However, the LDV sensitivity is much lower than that of conventional contact transducers. We therefore used a time-consuming signal averaging process to improve the signal-to-noise ratio. Consequently, clear images of the pulsed A_0 mode propagation waves were obtained as shown in Figure 4. On the other hand, there is a high demand for rapid non-destructive inspection techniques that can quickly indicate whether or not damage is present. Figure 5 depicts the visual results of ultrasonic wave propagation for the CFRP quasi-isotropic specimen without signal averaging. The obtained images are noisier than those in Figure 4b; however, the presence of the damage can still easily be identified through the reflected waves. Furthermore, we confirmed that non-contact and quick inspection is compatible with the impacted specimen used in this study.

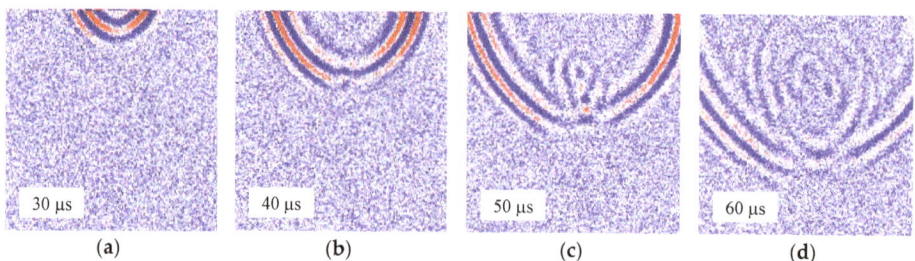

Figure 5. Lamb wave propagation in the impacted carbon fiber reinforced polymer (CFRP) quasi-isotropic laminate without signal averaging. (**a**) 30 µs; (**b**) 40 µs; (**c**) 50 µs; (**d**) 60 µs.

Figure 6 compares ultrasonic images at 55 µs for the CFRP quasi-isotropic specimen for various averaging times. As a result, the images become clearer with increasing averaging time (i.e., increasing scanning time). The averaging time should be selected according to the damage (type, size), objective materials, inspection time, inspection resolution, and so on. Moreover, the signal-to-noise ratio could be drastically improved and inspected quickly using the mid-infrared laser developed by Hatano et al. [24], as it can generate significantly larger ultrasonic amplitude in CFRP laminates than the conventional Nd: YAG laser [25].

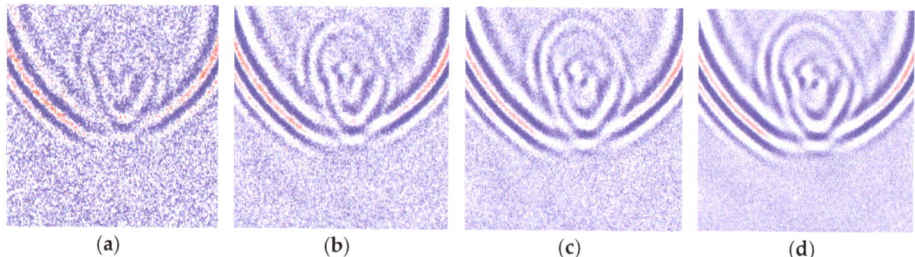

Figure 6. Effects of the averaging times on the ultrasonic images at 55 µs for carbon fiber reinforced polymer (CFRP) quasi-isotropic laminate. (**a**) non-averaged; (**b**) 5 times; (**c**) 10 times; (**d**) 30 times.

4. Discussion

To identify how the proposed non-contact method compares to the conventional methods [11–14], we analyzed the same specimen using a contact ultrasonic transducer instead of the LDV. Figure 7 depicts the visual results of the ultrasonic-wave propagation for the CFRP quasi-isotropic specimen using the conventional method. The contact ultrasonic receiver had a resonant frequency of 200 kHz (M204A, Fuji Ceramics, Fujinomiya, Japan) and was glued to the specimen surface at the position of the CW laser illumination. The received signals were amplified by a preamplifier (A1201, Fuji Ceramics, Fujinomiya, Japan) and bandpass-filtered from 50 to 400 kHz before being stored on the computer.

The sensitivity of the transducer is high enough for the S_0 mode to be observed. Similarly to the result shown in Figure 4, a phase delay is observed when the A_0 mode reaches the damage. However, the visualized incident wave of the A_0 mode is displayed as a continuous wave rather than a pulsed wave, causing the wave reflected from the damage to interact with the incident wave, obstructing its clear identification. Figure 8 compares the waveforms received by both methods when the pulsed laser illuminates the position shown by the black dot in Figure 7a. As expected, using the LDV unit, the incident wave of the A_0 mode has a pulsed shape; therefore, the wave reflected from the damage can be identified. In contrast, when using the contact transducer, the S_0 mode is observed and followed by a ringing phenomenon in the A_0 mode due to the resonance of the piezoelectric transducer. Due to the ringing, the reflected wave cannot be identified. This ringing often causes difficulty in detecting reflected waves from small defects or damage, and additional filtering becomes necessary for detailed analysis of the damage. In contrast, the LDV purely measures the out-of-plane displacement of the pulsed ultrasonic wave and thus does not cause such ringing. It should be noted that this is another significant advantage of the proposed non-contact method for easy inspection.

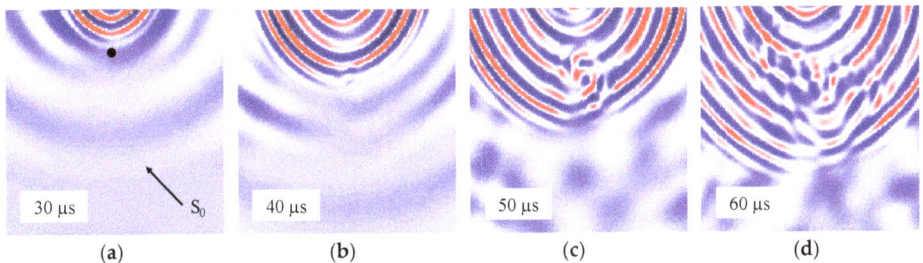

Figure 7. Lamb wave propagation in the impacted carbon fiber reinforced polymer (CFRP) quasi-isotropic laminate using a contact resonant ultrasonic transducer. (a) 30 µs; (b) 40 µs; (c) 50 µs; (d) 60 µs.

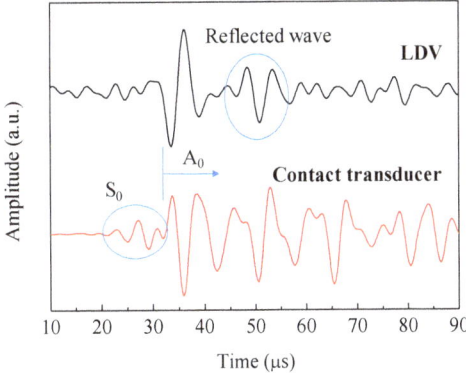

Figure 8. Comparison of the waveforms received by the laser Doppler vibrometer (LDV) and the contact transducer when pulsed laser illuminated at the black dot in Figure 7a.

5. Conclusions

This study demonstrated a rapid non-contact ultrasonic inspection technique by visualization of Lamb wave propagation for detecting barely visible impact damage in CFRP laminates. Our laser ultrasonic imaging system consists of a rapid pulsed laser scanning unit for ultrasonic generation and a LDV unit for ultrasonic reception. The signal-to-noise ratio of the ultrasonic signal was improved using retroreflective tape and a signal averaging process, and we successfully visualized the propagation of the pulsed Lamb A_0 mode in CFRP laminates. The interactions between the Lamb waves and the impact damage were clearly observed and the damages were easily detected through changes in wave propagation. Furthermore, we demonstrated that damage could be rapidly detected without applying signal averaging. The proposed method using the LDV has significant advantages in detecting damage compare to the conventional resonant ultrasonic transducer method. However, further improvement of the signal-to-noise ratio is necessary to inspect large areas and to detect smaller defects. In addition, we plan in the near future to establish an ultrasonic wavefield image dataset and to develop an automated image analysis system for damage detection in composite structures using machine learning.

Author Contributions: Conceptualization, N.T.; investigation, N.T. and S.Y.; methodology, J.Y. and W.K.; writing—original draft preparation, N.T.; writing—review and editing, J.Y., W.K. and S.Y.

Funding: This article is based on results obtained from a project commissioned by the New Energy and Industrial Technology Development Organization (NEDO).

Conflicts of Interest: The authors declare no conflict of interest.

References

1. Guo, N.; Cawley, P. The interaction of Lamb waves with delaminations in composite laminates. *J. Acoust. Soc. Am.* **1993**, *94*, 2240–2246. [CrossRef]
2. Hayashi, T.; Kawashima, K. Multiple reflections of Lamb waves at a delamination. *Ultrasonics* **2002**, *40*, 193–197. [CrossRef]
3. Diamanti, K.; Hodgkinson, J.M.; Soutis, C. Detection of low-velocity impact damage in composite plates using Lamb waves. *Struct. Health Monit.* **2004**, *3*, 33–41. [CrossRef]
4. Toyama, N.; Takatsubo, J. Lamb wave method for quick inspection of impact-induced delamination in composite laminates. *Compos. Sci. Technol.* **2004**, *64*, 1293–1300. [CrossRef]
5. Su, Z.; Ye, L.; Lu, Y. Guided Lamb waves for identification of damage in composite structures: A review. *J. Sound Vib.* **2006**, *295*, 753–780. [CrossRef]
6. Ramadas, C.; Padiyar, J.; Balasubramaniam, K.; Joshi, M.; Krishnamurthy, C.V. Lamb wave based ultrasonic imaging of interface delamination in a composite T-joint. *NDT&E Int.* **2011**, *44*, 523–530. [CrossRef]
7. Liu, Z.; Yu, H.; Fan, J.; Hu, Y.; He, C.; Wu, B. Baseline-free delamination inspection in composite plates by synthesizing non-contact air-coupled Lamb wave scan method and virtual time reversal algorithm. *Smart Mater. Struct.* **2015**, *24*, 045014. [CrossRef]
8. Feng, B.; Ribeiro, A.L.; Ramos, H.G. Interaction of Lamb waves with the edges of a delamination in CFRP composites and a reference-free localization method for delamination. *Measurement* **2018**, *122*, 424–431. [CrossRef]
9. Taheri, H.; Du, J.; Delfanian, F. Experimental observation of phased array guided wave application in composite materials. *Mater. Eval.* **2017**, *75*, 1308–1316.
10. Taheri, H. Utilization of Non-Destructive Testing (NDT) Methods for Composite Material Inspection (Phased Array Ultrasonic). Master's Thesis, South Dakota State University, Brookings, SD, USA, August 2014.
11. Takatsubo, J.; Wang, B.; Tsuda, H.; Toyama, N. Generation laser scanning method for the visualization of ultrasounds propagating on a 3-D object with an arbitrary shape. *J. Solid Mech. Mater. Eng.* **2007**, *1*, 1405–1411. [CrossRef]
12. Yashiro, S.; Takatsubo, J.; Miyauchi, H.; Toyama, N. A novel technique for visualizing ultrasonic waves in general solid media by pulsed laser scan. *NDT&E Int.* **2008**, *41*, 137–144. [CrossRef]
13. Yashiro, S.; Takatsubo, J.; Toyama, N. An NDT technique for composite structures using visualized Lamb-wave propagation. *Compos. Sci. Technol.* **2007**, *67*, 3202–3208. [CrossRef]
14. Toyama, N.; Yamamoto, T.; Urabe, K.; Tsuda, H. Ultrasonic inspection of adhesively bonded CFRP/aluminum joints using pulsed laser scanning. *Adv. Compos. Mater.* **2017**. [CrossRef]
15. Sohn, H.; Dutta, D.; Yang, J.Y.; Park, H.J.; DeSimio, M.; Olson, S.; Swenson, E. Delamination detection in composites through guided wave field image processing. *Compos. Sci. Technol.* **2011**, *71*, 1250–1256. [CrossRef]
16. Chia, C.C.; Lee, J.-R.; Park, C.-Y.; Jeong, H.-M. Laser ultrasonic anomalous wave propagation imaging method with adjacent wave subtraction: Application to actual damages in composite wing. *Opt. Laser Technol.* **2012**, *44*, 428–440. [CrossRef]

17. Lee, J.-R.; Chia, C.C.; Park, C.-Y.; Jeong, H. Laser ultrasonic anomalous wave propagation imaging method with adjacent wave subtraction: Algorithm. *Opt. Laser Technol.* **2012**, *44*, 1507–1515. [CrossRef]
18. Michaels, T.E.; Michaels, J.E.; Ruzzene, M. Frequency–wavenumber domain analysis of guided wavefields. *Ultrasonics* **2011**, *51*, 452–466. [CrossRef]
19. Rogge, M.D.; Leckey, C.A. Characterization of impact damage in composite laminates using guided wavefield imaging and local wavenumber domain analysis. *Ultrasonics* **2013**, *53*, 1217–1226. [CrossRef]
20. Park, B.; An, Y.-K.; Sohn, H. Visualization of hidden delamination and debonding in composites through noncontact laser ultrasonic scanning. *Compos. Sci. Technol.* **2014**, *100*, 10–18. [CrossRef]
21. An, Y.-K. Impact-induced delamination detection of composites based on laser ultrasonic zero-lag cross-correlation imaging. *Adv. Mater. Sci. Eng.* **2016**, *2016*, 6474852. [CrossRef]
22. Kudela, P.; Radzienski, M.; Ostachowicz, W. Impact induced damage assessment by means of Lamb wave image processing. *Mech. Syst. Signal Pr.* **2018**, *102*, 23–36. [CrossRef]
23. Pavlakovic, B.; Lowe, M.; Alleyne, D.; Cawley, P. Disperse: A general purpose program for creating dispersion curves. In *Review of Progress in Quantitative Nondestructive Evaluation*; Thompson, D.O., Chimenti, D.E., Eds.; Springer: Boston, MA, USA, 1997; Volume 16, pp. 185–192, ISBN 9781461377252.
24. Hatano, H.; Watanabe, M.; Kitamura, K.; Naito, M.; Yamawaki, H.; Slater, R. Mid IR pulsed light source for laser ultrasonic testing of ca-bon-fiber-reinforced plastic. *J. Opt.* **2015**, *17*, 094011. [CrossRef]
25. Kusano, M.; Hatano, H.; Watanabe, M.; Takekawa, S.; Yamawaki, H.; Oguchi, K.; Enoki, M. Mid-infrared pulsed laser ultrasonic testing for carbon fiber reinforced plastics. *Ultrasonics* **2018**, *84*, 310–318. [CrossRef] [PubMed]

© 2018 by the authors. Licensee MDPI, Basel, Switzerland. This article is an open access article distributed under the terms and conditions of the Creative Commons Attribution (CC BY) license (http://creativecommons.org/licenses/by/4.0/).

Article

In Situ Analysis of Plaster Detachment by Impact Tests

Alessandro Grazzini

Department of Structural Geotechnical and Building Engineering, Politecnico di Torino, 10129 Torino, Italy; alessandro.grazzini@polito.it

Received: 18 December 2018; Accepted: 9 January 2019; Published: 12 January 2019

Abstract: The frescoed surfaces of historical buildings may be subject to detachment due to various causes of deterioration. A new non-destructive experimental methodology is described to assess in situ the safety against plaster detachments from historical wall surfaces. Through small and punctual impacts exerted with a specific hammer on the plastered surface it is possible to evaluate the level of the plaster's detachment. A case study at Palazzo Birago in Turin (Italy) is described to give an example of the application of this innovative technique on frescoed surfaces of historical vaults. The test allows to evaluate the safety of frescoed decorations without affecting the material consistency or creating damage, therefore, making it very suitable in the field of architectural heritage.

Keywords: frescoed surfaces; non-destructive test; plaster detachment; impact hammer test; historical masonry building

1. Introduction

In the field of historical buildings, the role of monitoring and diagnostics is increasingly important for the purpose of securing the masonry structures and also the decorative apparatuses. Often the normal degradation over time or the external climatic causes can compromise the stability of historical plasters [1,2]. The potential detachment of plaster can be further dangerous if it comes from masonry vaults, with the risk of material inside historical buildings containing residential or public functions falling. The Non-Destructive Testing Laboratory of the Politecnico di Torino introduced an impact method to be applied on the wall surface. By means of an instrumented hammer, the impact of a known mass with predetermined energy against the plaster surface was produced. The force–time diagrams produced by the impact of a mass of known energy against the test surface were analyzed. The knowledge of the evolution over time of the forces between the impact mass and the tested material allows to evaluate the parameters that other impulsive methods would not allow. For example, the concrete sclerometer test was limited to the detection of a single quantity, i.e., the elastic energy returned by the material after the impact, proportional to the rebound length of the mass. In the case of the impact hammer test, in addition to the elastic energy returned by the material, the given energy, the dissipated energy, the duration of the impact, and the maximum force can also be evaluated. The impact method allowed to assess the elastic and anelastic properties of the materials [3].

An experimental analysis of the stability of the decorated plaster covering three masonry vaults of the Birago Palace (16th century, planned by Filippo Juvarra) in the center of Turin was described by the use of the impact method test. The frescoed vaults of the Pelagi, Blu, and Giunta rooms showed some small cracks branched out in a layer of plaster that needed an evaluation regarding the risk of detachment (Figures 1–3). Surveys carried out on several points of the vaulted surfaces allowed mapping of the points of potential detachment of the de-coated plaster.

Figure 1. (**a**) Pelagi room at Birago Palace; (**b**) cracks branched in the decorated plaster of the masonry vault in the Pelagi room.

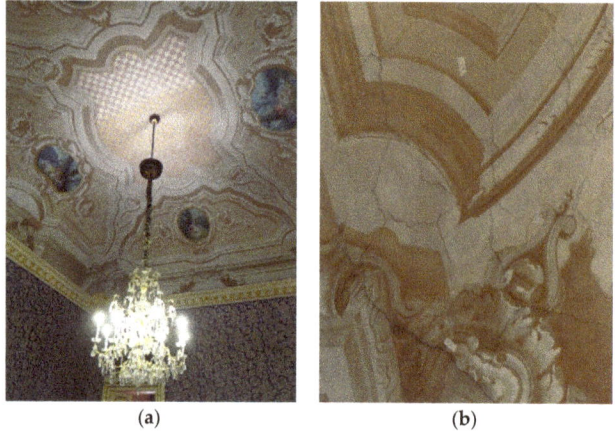

Figure 2. (**a**) Blu room at Birago Palace; (**b**) cracks branched in the decorated plaster of the masonry vault in the Blu room.

Figure 3. (**a**) Giunta room at Birago Palace; (**b**) cracks branched in the decorated plaster of the masonry vault in the Giunta room.

2. Equipment Setup and Methods

The instrumentation used to carry out the tests consisted of an impact instrumented hammer and a data analyzer. The electric impact hammer used was: PCB Piezotronics; model 086B09; force variable from 44.48 N to 4448.26 N; 208M51 model PCB force sensor; force sensor sensitivity 2.47 mV/N

(Figure 4a). The electric impact hammer was predetermined energy, characterized by the presence of one amplifier level and impedance adapter and a spherical head (10 mm diameter) in cemented steel rigidly connected to a piezoelectric impulse transducer with a total mass of 207 g.

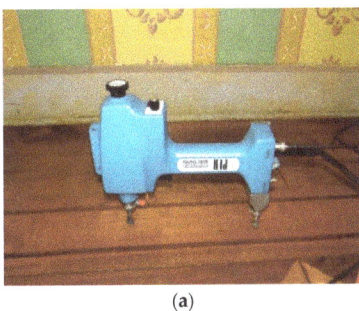 

(a) (b)

Figure 4. (a) The electric impact hammer; (b) the LMS Pimento multi-channel signal analyzer.

The LMS Pimento multi-channel signal analyzer, with the "real-time" acquisition and recording function, had the following characteristics: model MSP 424; number of channels 4; input range: ± 316 mV $\div \pm 31.6$ mV; 24-bit ADC (analog digital converter); bandwidth greater than 20 kHz (on all channels); signal sampling rate up to 100 ksample/second; and personal computer interface: FireWire IEEE1394—managed by its own dedicated software (Figure 4b).

The points of the vault were randomly selected according to the logistic possibilities of movement inside the three rooms through mobile scaffolding. Most of the points chosen were found inside the cracked areas where there was a need to evaluate the adherence of the plaster on the masonry vault. Several points were also analyzed in non-cracked areas in order to compare the experimental results with the cracked points. For each point, at least three acquisitions were made to improve the statistical data (Figure 5a). For each single point test, the instrumented hammer was positioned with the impact mass perpendicular to the test surface. The perpendicularity was achieved by means of the four metallic footsies (Figure 5b). The test consisted in generating a small impact of the hammer's mass against the test surface, with an absolutely non-destructive intensity, and therefore, also compatible with the conservation of the frescoed surfaces. The impact was triggered by a trigger control on the electric impact hammer.

(a) (b)

Figure 5. (a) Use of the electric impact hammer for the adherence test of the frescoed plaster of the masonry vaults at Birago Palace; (b) calibration test where it is possible to see the positioning of the electric impact hammer on the test surface.

3. Impact Energy Principles

The following are some energy considerations to better understand the theory underlying the impact method. Consider the impact of a mass m with a semispherical surface and having a velocity v_0 on the flat surface of a semi-finished space. The direction of impact is perpendicular to this surface. Moreover, the velocity v_0 of all points of the mass is equal and coinciding with the velocity v_0 of its center of gravity. In this case the kinetic energy of the mass at the moment of impact is given by Equation (1):

$$\varepsilon_{c1} = \frac{1}{2}mv_0^2 \tag{1}$$

and the momentum is:

$$Q_1 = mv_0. \tag{2}$$

Considering the instant t_0 in which the mass touches the surface and the instant t_1 in which the maximum contact deformation δ occurs and in which the velocity v_0 is canceled (Figure 6a), the corresponding momentum variation results:

$$|mv_0 - mv_1| = \int_{t_0}^{t_1} F dt \tag{3}$$

and therefore:

$$mv_0 = \int_{t_0}^{t_1} F dt \tag{4}$$

The force impulse is given by the area A_1 subtended to the curve (F, t) obtained experimentally as shown in the Figure 6b.

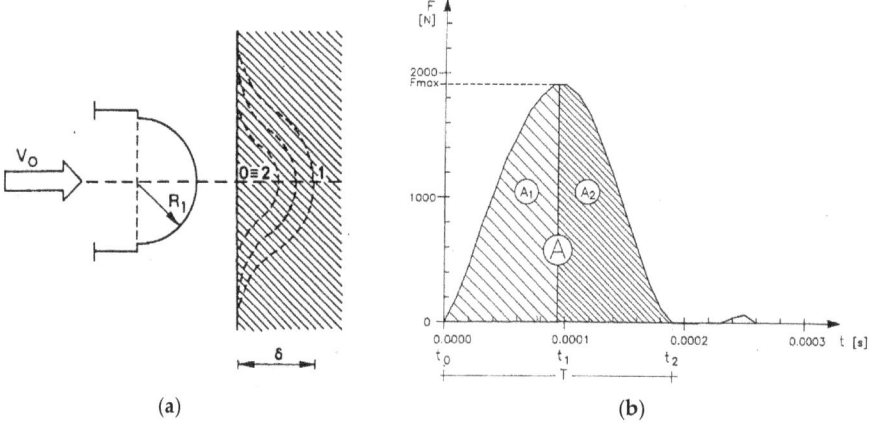

(a)   (b)

Figure 6. (a) Geometry of the mobile mass; (b) force–time curve obtained from an impact test.

The kinetic energy provided by the mass is:

$$\varepsilon_{c1} = \frac{1}{2}mv_0^2 = \frac{\left(\int_{t_0}^{t_1} F dt\right)^2}{2m} \tag{5}$$

Subsequently from the instant t_1, in which the vector displacement of the mass changes direction, at the instant t_2, in which the contact between mass and flat surface ceases, the change in momentum of the mass results:

$$mv_2 = \int_{t_1}^{t_2} F dt \tag{6}$$

wherein v_2 is the velocity of displacement of the mass from the surface, and results $v_2 < v_0$. The value of the integral (6) is given by the area A_2. The ratio between initial and final mass momentum provides the return coefficient e that measures the elasticity of the impact:

$$\frac{mv_2}{mv_0} = \frac{v_2}{v_0} = e \tag{7}$$

In the perfectly elastic collision $e = 1$, in the perfectly inelastic collision $e = 0$.
Figure 6b shows that the return coefficient is given by the following equation:

$$\frac{\int_{t_1}^{t_2} F dt}{\int_{t_0}^{t_1} F dt} = \frac{A_2}{A_1} = e \tag{8}$$

The energy returned in the collision is given by:

$$\varepsilon_{c2} = \frac{1}{2} m v_2^2 = \frac{\left(\int_{t_1}^{t_2} F dt\right)^2}{2m} \tag{9}$$

The ratio between the energy supplied and returned is given by:

$$\frac{\varepsilon_{c2}}{\varepsilon_{c1}} = e^2 \tag{10}$$

The energy dissipated ε_d in the impact due to the elasticity of the materials is given by:

$$\varepsilon_d = \left(1 - e^2\right) \varepsilon_{c1} \tag{11}$$

Therefore, in the case of a perfectly elastic impact:

$$e = 1, \text{ i.e., } A_2 = A_1 \tag{12}$$

4. Experimental Results at Birago Palace Tests

In the impact test carried out at Birago Palace in Turin (Italy), every masonry vault was divided into survey areas as shown in Figure 7, labeled with alphabet letters, within which both apparently intact and potentially damaged points were tested. For each point, at least three impacts were performed to obtain a better statistical response, and the return coefficient e was evaluated. The maps of the areas tested and the force–time curves of some tested points are shown, respectively, in the Pelagi (Figures 8 and 9), Blu (Figures 10 and 11), and Giunta rooms (Figures 12 and 13). Tables 1–3 show the coefficient averages for each point.

Figure 7. Survey area on the Pelagi room vault.

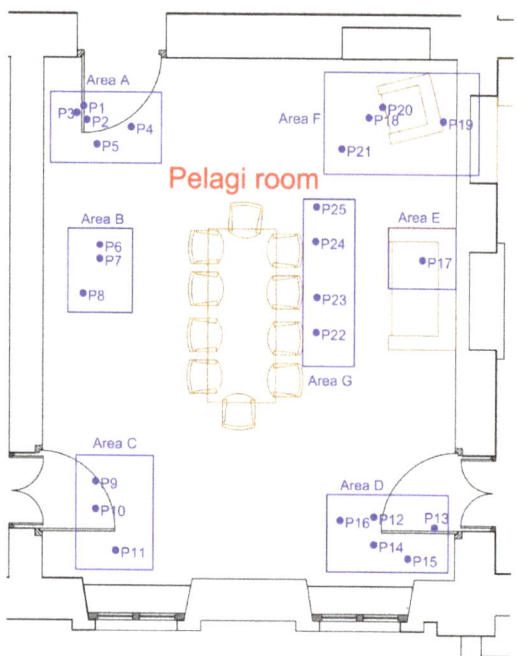

Figure 8. Map of the points and areas tested on the vault of the Pelagi room.

Table 1. Results of impact test on Pelagi room vault.

Area	Test Point	$e = A_2/A_1$	Notes	Adhesion Plaster
A	P1	0.77		safe
	P2	1.21	stuccoing	not safe
	P3	0.83		safe
	P4	1.08		not safe
	P5	1.33	stuccoing	not safe
B	P6	0.74		safe
	P7	0.73		safe
	P8	0.71		safe
C	P9	0.66		safe
	P10	0.91		safe
	P11	1.36		not safe
D	P12	0.66		safe
	P13	0.62		safe
	P14	0.65		safe
	P15	0.73		safe
	P16	0.75		safe
E	P17	0.91		safe
F	P18	0.69		safe
	P19	0.73		safe
	P20	1.43		not safe
	P21	0.68		safe
G	P22	0.60		safe
	P23	0.62		safe
	P24	1.19		not safe
	P25	0.69		safe

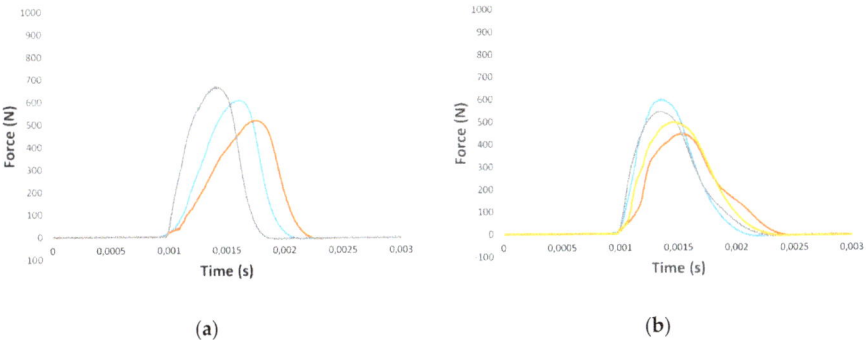

(a)　(b)

Figure 9. Impact test on Pelagi room vault: force–time curve of (**a**) P13 test; (**b**) P5 test.

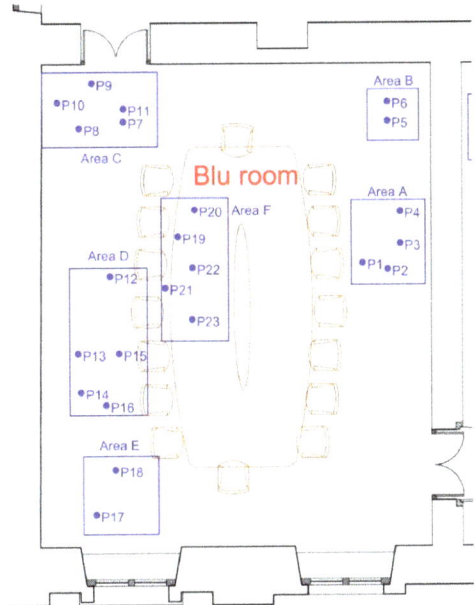

Figure 10. Map of the points and areas tested on the vault of the Blu room.

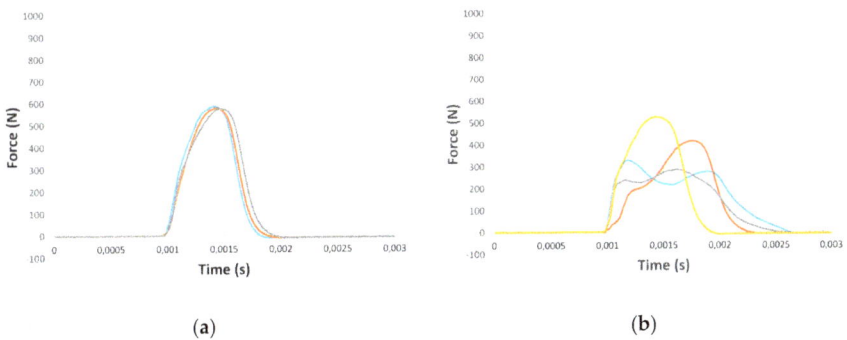

(a)　(b)

Figure 11. Impact test on Blu room vault: Force–time curve of (**a**) P7 test; (**b**) P4 test.

Table 2. Results of impact test on Blu room vault.

Area	Test Point	$e = A_2/A_1$	Notes	Adhesion Plaster
A	P1	0.68		safe
	P2	0.75		safe
	P3	0.71		safe
	P4	1.98	Stuccoing	not safe
B	P5	0.78		safe
	P6	0.65		safe
C	P7	0.77		safe
	P8	0.76		safe
	P9	0.84		safe
	P10	0.79		safe
	P11	0.78		safe
D	P12	0.75		safe
	P13	1.11		not safe
	P14	0.87		safe
	P15	0.70		safe
	P16	0.56		safe
E	P17	0.83		safe
	P18	0.63		safe
F	P19	0.72		safe
	P20	0.66		safe
	P21	0.66		safe
	P22	0.70		safe
	P23	0.73		safe

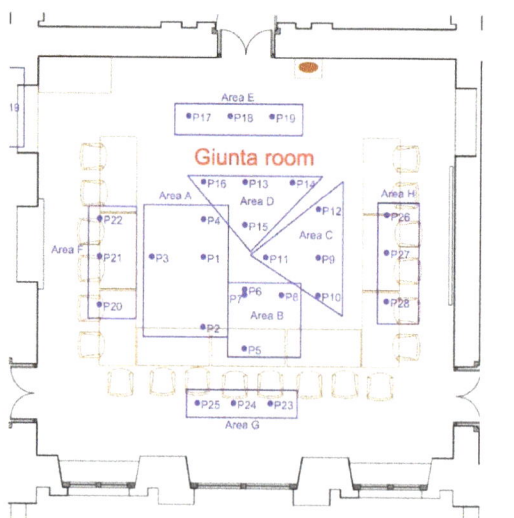

Figure 12. Map of the points and areas tested on the vault of the Giunta room.

It is possible to observe that most of the points tested had a return coefficient e lower than 1. This means that the energy returned was lower than that emitted, because part of this energy was dissipated by the tested structure through sufficient bonds in the interface between the plaster and the masonry surface. On the contrary, the return coefficient $e > 1$ showed a returned energy greater than the one emitted: in this case the material was already damaged [4,5] because it was partly or completely disconnected and returned more energy due to the deformations and the microscopic movements active due the non-perfect adherence between plaster and masonry surface.

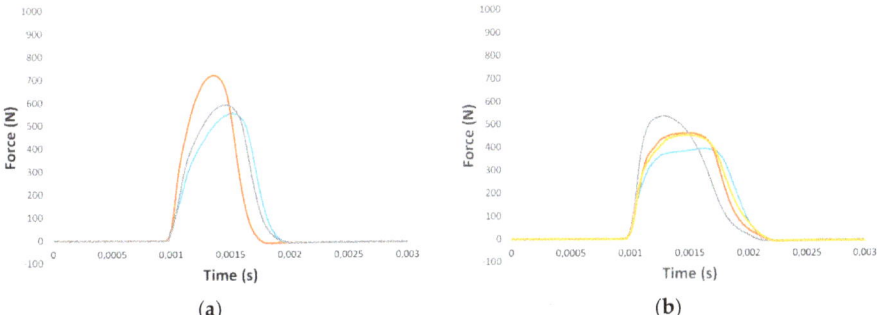

Figure 13. Impact test on Giunta room vault: force–time curve of (a) P4 test; (b) P15 test.

Table 3. Results of impact test on Giunta room vault.

Area	Test Point	$e = A_2/A_1$	Notes	Adhesion Plaster
A	P1	0.68		safe
	P2	0.74		safe
	P3	0.70		safe
	P4	0.65		safe
B	P5	0.84		safe
	P6	0.73		safe
	P7	0.72		safe
	P8	0.77		safe
C	P9	0.93		safe
	P10	0.70		safe
	P11	0.80		safe
	P12	0.72		safe
D	P13	0.65		safe
	P14	0.78		safe
	P15	1.07	Wiring channel	not safe
	P16	0.74		safe
E	P17	1.04		not safe
	P18	1.26		not safe
	P19	0.70	Stuccoing	safe
F	P20	0.74		safe
	P21	0.74		safe
	P22	0.68		safe
G	P23	0.66		safe
	P24	0.62		safe
	P25	0.77		safe
H	P26	0.75		safe
	P27	0.71		safe
	P28	0.62		safe

5. Discussion

The points where the plaster was still adherent to the wall surface showed more symmetrical and regular force–time curves, with higher values than the maximum impact force as the material was more compact (Figures 9a, 11a and 13a). On the contrary, in the points already covered by previous stuccoing, lower values of the maximum force and more asymmetric curves were recorded in which the area after maximum force was greater than that which preceded it (Figures 9b, 11b and 13b).

Overall, the impact test showed the stability and safety of the adhesion between decorated plaster and masonry surfaces of the vaults examined in the three rooms. Some points of lesser safety regarding adherence have emerged. Some of these concerns point to previously stuccoed areas only a few decades old. This point highlighted the potential critical stability of some of the plaster, which had already been the subject of micro-grouting, and for which restorations had not been perfectly carried out. On the contrary many other previously stuccoed points showed a return coefficient <1. The impact method was therefore also useful to qualify the effectiveness of previous restoration work.

On the masonry vault of the rooms some points of potential detachment of plaster have been found, characterized by a return coefficient $e > 1$ (Tables 1–3). These results were in agreement with what has been possible to perceive qualitatively with a simple hand knock on the point under investigation. Some of these points with a high return coefficient were stuccoed previously, a sign that some critical issues of potential detachment already existed in the past (Figure 14). In some cases, as for point P15 of the Giunta room, the impact method confirmed the presence of an installation channel that feeds the chandelier as a zone with weak adherence of the plaster (Figures 13b and 14b).

Figure 14. Points of potential plaster detachment: (**a**) P11 vault Pelagi room; (**b**) P15 vault Giunta room.

The decorated surfaces of the vaults therefore appeared to be in a good state of conservation. The ramified cracks present in most of the surface derive from the shrinkage effects of the historical plaster mortar, due to different reasons: climatic conditions, setting of the binders (lime and cement), binder/inert quantity ratio.

6. Conclusions

The impact method was used to evaluate the adherence of the decorated plaster of some masonry vaults. The method confirmed its non-destructive typology and has proved its validity also for the diagnostics of historical buildings. In the campaign tests carried out to evaluate the adherence of the decorated plaster on three masonry vaults of the Birago Palace (Turin, Italy), the impact method clearly highlighted points of critical and potential detachment, as well as confirmed the effectiveness of the consolidation of many previously stuccoed points.

Funding: This research was funded by the CAMERA DI COMMERCIO DI TORINO.

Acknowledgments: The author wishes to thank Vincenzo Di Vasto for his valuable collaboration during the performance of the tests.

Conflicts of Interest: The authors declare no conflict of interest.

References

1. Bocca, P.; Valente, S.; Grazzini, A.; Alberto, A. Detachment analysis of dehumidified repair mortars applied to historical masonry walls. *Int. J. Arch. Herit.* **2014**, *8*, 336–348. [CrossRef]

2. Grazzini, A.; Lacidogna, G.; Valente, S.; Accornero, F. Delamination of plasters applied to historical masonry walls: Analysis by acoustic emission technique an numerical model. *IOP Conf. Ser. Mater. Sci. Eng.* **2018**, *372*, 1–7. [CrossRef]
3. Bocca, P.; Scavia, C. The impulse method for the evaluation of concrete elastc characteristics. In Proceedings of the 9th International Conference on Experimental Mechanics, Copenhagen, Denmark, 20–24 August 1990.
4. Bocca, P.; Carpinteri, A.; Valente, S. On the applicability of fracture mechanics to masonry. In Proceedings of the 8th International Brick/Block Masonry Conference, Dublin, Ireland, 19–21 September 1988.
5. Johnson, K.L. *Contact Mechanics*; Cambridge University Press: Cambridge, UK, 1985.

© 2019 by the author. Licensee MDPI, Basel, Switzerland. This article is an open access article distributed under the terms and conditions of the Creative Commons Attribution (CC BY) license (http://creativecommons.org/licenses/by/4.0/).

Article

A Nonlinear Method for Characterizing Discrete Defects in Thick Multilayer Composites

Guoyang Teng [1], Xiaojun Zhou [1], Chenlong Yang [1,*] and Xiang Zeng [2]

[1] The State Key Lab of Fluid Power and Mechatronic Systems, Zhejiang University, Hangzhou 310027, China; t_gy189@163.com (G.T.); cmeesky@163.com (X.Z.)
[2] CRRC Zhuzhou Institute Co., Ltd., Zhuzhou 412001, China; zzjjuu0104@163.com
* Correspondence: zjuppt@163.com; Tel.: +86-157-0007-9582

Received: 1 February 2019; Accepted: 15 March 2019; Published: 20 March 2019

Abstract: Discrete defects in thick composites are difficult to detect for the small size and the structure noise that appears in multilayer composites. In this paper, a nonlinear method, called recurrence analysis, has been used for characterizing discrete defects in thick section Carbon Fiber Reinforced Polymer (CFRP) with complex lay-up. A 10 mm thick CFRP specimen with nearly zero porosity was selected, and blind holes with different diameters were artificially constructed in the specimen. The second half of the backscattered signal was analyzed by recurrence analysis for areas with or without a defect. The recurrence plot (RP) visualized the chaotic behavior of the ultrasonic pulse, and the statistical results of recurrence quantification analysis (RQA) characterized the instability of the signal and the effect of defects. The results show that the RQA variable differences are related to the size of blind holes, which give a probable detection of discrete geometric changes in thick multilayer composites.

Keywords: thick multilayer composites; discrete defects; ultrasonic pulse echo; nondestructive testing (NDT); recurrence plot (RP); recurrence quantification analysis (RQA); statistical results; chaotic behavior

1. Introduction

Carbon Fiber Reinforced Polymer (CFRP) is one of the most widely used multilayer composites in aerospace due to its specific features, such as high ratio of strength to weight, high modulus, and high fatigue resistance. Discrete defects, like larger voids, delaminations, and cracks can occur during manufacturing or service process, and they may result in a significant loss of mechanical properties [1]. Thus, early defect detection is essential to avoid serious problems that are caused by defects. Nondestructive testing (NDT) has been employed to characterize discrete defects in composite structures for many years [2,3]. As an important NDT method, ultrasonic testing has been widely used in the evaluation of defects in CFRP. Li et al. [4] used the ultrasonic arrays technique to improve the characterization of side drilled holes in a 19 mm thick CFRP block, in which the holes were 1.5 mm in diameter and 16 mm in length and down to a depth of 16 mm. Ibrahim et al. [5] performed single-sided technique of contact pulse-echo inspection on CFRP specimen with thickness of 10 mm, to study the effect of crack in the middle of the specimen, while the length of the crack is approximately 25 mm. Smith et al. [6] have successfully applied a two-dimensional fast Fourier transform (2D-FFT) method to B-scan images for detecting out of plane fiber waviness in structures that are as thick as 18 mm. Most researches of inspecting thick CFRP were focused on defects with large sizes, however defects, such as larger voids or micro-cracks, may be smaller than 1 mm. It is not yet able to specify limitations for discrete defects in thick section CFRP. More reliable and quantitative studies are required [3].

Unlike the signals of metallic materials, there are no clearly identifiable defect echoes in the signal of multilayer composites, since the fiber layer with a different direction may also cause a reflected ultrasonic echo [7]. Although ultrasonic inspection can be performed in a variety of physical configurations, for example, the ultrasonic array, pulse-echo inspection is most straightforward and practical among all of the ultrasonic techniques. The detection of discrete defects is mainly based on signal processing of the backscattered signal, which is between the front surface echo and the back wall echo in a typical ultrasonic signal, as shown in Figure 1. Due to the layered structure with a layer thickness that is close to the wavelength of the ultrasonic pulse, the backscattered signal will exhibit resonance noise that is continuously attenuated. Dominguez [8] believes that the frequency continuity and amplitude degradation of backscattered signals will be destroyed for local defects, which can be detected by time-frequency analysis and time-energy analysis. This method is effective for thin CFRP. However, as the thickness of the material increases, defects may occur in the portion where the resonance noise has been attenuated, and the defect echoes are mixed with echoes that are caused by the material structure. The amplitude and energy of the backscattered signal are low, while the signal-to-noise ratio (SNR) of the defect is also low. Thus, the defect echo in the second half of the backscattered signal of the thick section CFRP cannot be well distinguished by traditional time-frequency analysis or time-energy analysis.

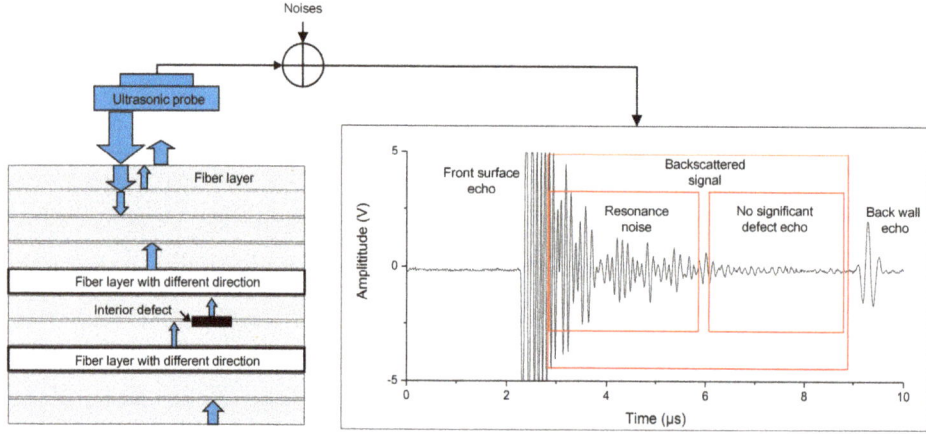

Figure 1. Schematic diagram of the pulse-echo method and an A-scan signal of Carbon Fiber Reinforced Polymer (CFRP) with interior defects and layers with different direction.

In order to solve the problem of characterizing defects in composites, model driven methods, such as ultrasonic pulse-echo modeling and structural modelling of composites, have been researched. A generalized parametric ultrasonic echo model and algorithms for accurately estimating the parameters have been presented in literature [9]. In Part II of the study, the advantage of the model-based estimation method in ultrasonic applications has been explored [10]. Using model-based methods, the ultrasonic signal waveform consisting of multiple overlapping echoes from within thin multilayer structures has been successfully reconstructed [11]. However, thick multilayer composites have large number of layers and diverse layering methods, and they are difficult to be described by a generic model. A data processing method that is capable of dealing with nonlinearity in ultrasound signals would be more useful and suitable for thick multilayer composites for now.

Recurrence analysis has been applied in a wide range of fields, including weather analysis, biological medicine, economic analysis, signal processing, and so on [12,13]. Recently, recurrence analysis has proven to be useful in the ultrasonic testing of porous materials. Using recurrence quantification analysis (RQA), Carrión A. et al. [14] propose the ultrasonic signal modality as a new approach for damage evaluation in concrete, and the results show that one of the RQA variables is more sensitive to damage in spoiled series than other NDT techniques. They also adopt recurrence analysis for the characterization of scattering material with different porosity and propose the measurement of predictability as an indicator of percentages of porosity [15]. Besides, Brandt has used RQA in ultrasonic testing of CFRP [16,17] for the assessment of porosity. The works focus on structures, where the evaluation of the back wall echo from the opposite side of the ultrasonic probe is not able to made, and try to find the relationship between RQA variables and porosity to get an equivalent back wall echo.

For porosity that is distributed throughout the volume, recurrence analysis of the entire time series yields good results, while discrete defects are locally distributed in the composite. The nonlinearity of ultrasonic pulses has been proven to be sensitive to distributed voids, while the recurrence analysis method may also be useful in characterizing the nonlinearity of discrete defects and needs to be experimentally studied. Therefore, in this paper, a 10 mm thick CFRP specimen with 80 layers and zero porosity was tested while using the ultrasonic pulse echo method. Blind holes with different diameters that are smaller than or equal to 1 mm at depth of 6 mm were artificially conducted in the specimen and the signals were analyzed by the recurrence analysis method. The statistics of RQA variables were used to characterize the stability of backscattered signals for inspection of discrete defects with small sizes.

2. Methodology

If the data is aperiodic and does not recognize simple rules of their time dependence, then an approximate repetition of certain events, called a recurrence, can help us build more complex rules [18]. The recurrence analysis of the dynamics of a system is conducted in a phase space that was constructed with delayed vectors. A sequence of scalar measurements $x(t_n)$, $n = 1, 2, \ldots, N$ can be extended to a vector by the Takens delay method [19]:

$$\vec{x}_n = (x_{n-(m-1)\tau}, x_{n-(m-2)\tau}, \cdots, x_{n-\tau}, x_n) \tag{1}$$

where τ is the time lag and m is the embedding dimension. The values of m and τ determine the fact of whether the required information can be obtained from the original time series, while improper parameters will seriously distort the analysis seriously. The common selection method of embedded dimension m is based on false nearest neighborhood, and the selection method of delay time τ is the average mutual information method, according to reference [20].

A method for visualizing recurrences is called a recurrence plot (RP) and Eckmann et al. have introduced it [21]. Compute the matrix

$$R_{i,j} = \Theta(\varepsilon - \|x_i - x_j\|), i, j = 1, 2, \ldots, n - (m - 1) \cdot \tau \tag{2}$$

where Θ is the Heaviside step function (i.e., $\Theta(x) = 0$ if $x < 0$, and $\Theta(x) = 1$ otherwise), $\|\bullet\|$ is the Euclidean distance between the two vectors, ε is a tolerance parameter to be chosen, and x_i is the delayed vectors of some embedding dimension. Darken all of the nonzero values in the recurrence matrix $R_{i,j}$ and the RP is attained, as shown in Figure 2, in which many special structures exist. According to the macroscopic structures of the RP, the characteristics of plots refer to the different dynamics of system. Figure 2a is a homogeneous RP, which is shown as a uniform distribution of single recurrence dots, and it represents a typical stationary system, such as a random time series. Figure 2b is a period RP, which is shown as a long diagonal structure, and it represents a periodic oscillation system. Figure 2c is the RP of a chaotic system, shorter diagonal lines, small blocks, and single dots can be found as the suggest of chaos [22].

There are also microscopic features, such as single dots, diagonal lines, and vertical and horizontal lines in the RP. The appearance of single dots indicates that the corresponding state does not last or greatly fluctuates. Diagonal lines consist of a series of adjacent recurrence dots, and most of them are parallel to the main diagonal line. A diagonal line represent that the system track is similar in the same direction within a certain time period, and its length represents the degree of determination or predictability. Vertical or horizontal lines represent time segments that remain unchanged or change very slowly, and they are typical behaviors of the state of the laminate, which can reveal the discontinuity of the signal.

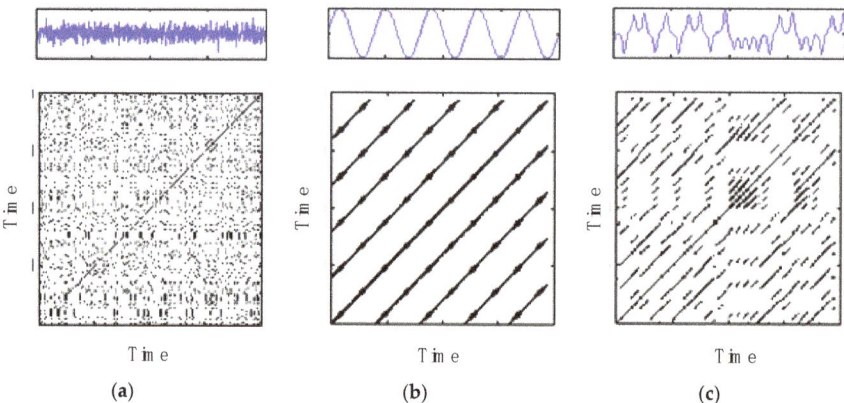

Figure 2. Different types of signals and their RPs. (**a**) white noise; (**b**) a periodic signal (cosine wave); and, (**c**) a chaotic system (the Lorenz system).

Usually, the macrostructure of RP can help us to directly observe the differences in the general structure of the system, while the results are significantly affected by the individual subjective judgment. Therefore, it is necessary to conduct a quantitative analysis that is based on microstructure. The statistical method RQA is more persuasive. In general, RP analysis provides a visual inspection of the matrix in Equation (2), and RQA analysis provides statistical variables that are based on diagonal, vertical, or horizontal lines formed by recurrence dots in the matrix.

Based on the diagonal lines in the RP, Zbilut and Webber put forward some quantities to measure the complexity of the system [23]:

1. Recurrence rate (RR)

$$RR(\varepsilon) = \frac{1}{N^2 - N} \sum_{i \neq j=1}^{N} R_{i,j}(\varepsilon) \qquad (3)$$

counts the black dots in the RP excluding the main diagonal line. RR is a measure of the relative density of recurrence points in the recurrence matrix.

2. RQA variables that are based on diagonal lines

$$DET = \frac{\sum_{l=l_{min}}^{N} l H_D(l)}{\sum_{i,j=1}^{N} R_{i,j}} \qquad (4)$$

Percent determinism (DET), the ratio of recurrence points that form diagonal lines to all recurrence points, while the length of diagonal lines should be larger than l_{min}. Usually, $l_{min} = 2$. $H_D(l)$ is the histogram of the lengths of the diagonal structures in the RP.

There are not only diagonal lines in RP, but also vertical and horizontal line segments. From these structures, Marwan et al. [20] proposed extended recurrence quantization variables:

3. RQA variables that are based on vertical and horizontal lines

$$LAM = \frac{\sum\limits_{l=v_{min}}^{N} lH_V(l)}{\sum\limits_{i,j=1}^{N} R_{i,j}} \quad (5)$$

The definition of the laminarity (LAM) is similar to the definition of DET and it represents the percentage of recurrence points in vertical structures. Analogously, $H_V(l)$ is the histogram of lengths of vertical lines with v_{min} as the minimal length of vertical lines in RP. Usually $v_{min} = 2$.

Figure 3 shows the overall framework of recurrence analysis.

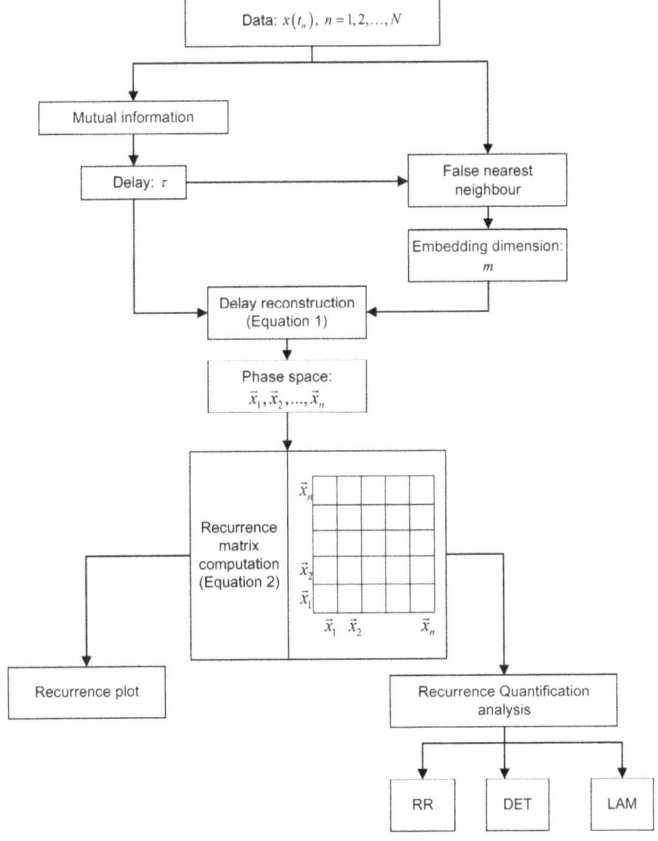

Figure 3. The framework of recurrence analysis.

3. Experiment

3.1. Material and Test Set-Up

The ultrasonic measurement device is mainly composed of an industrial personal computer, an ultrasonic acquisition card, an ultrasonic probe, and a set of position adjustment mechanism. Figure 4 shows the system. The ultrasonic probe is the OLYMPUS immersion plane probe (I3-0708-R, Resolution Series) with a center frequency of 7.5 MHz. The ultrasonic acquisition card model is

PCIUT3100 and the card can achieve the function of ultrasonic pulse transmission/reception at a sampling rate of 100 MHz. The IPC is ADVANTECH IPC-6608. Once the specimen is flattened to the fixture, the adjustment mechanism is used to adjust the vertical position of the probe to make the ultrasonic waveform clear. Subsequently, the vertical position of the probe should remain the same, while the horizontal position of the probe adjusted while using the adjustment mechanism to detect different areas of the specimen.

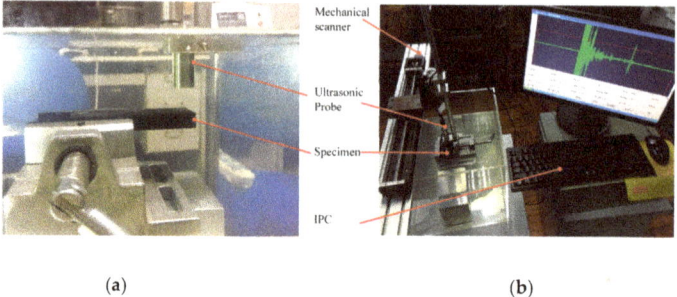

(a) (b)

Figure 4. Ultrasonic testing system (a) zoom of the probe onto scanned specimen (b) the whole testing system.

The experimental material is a thick section CFRP specimen that is provided by an aircraft manufacturing company. The reinforcement is carbon fiber and the matrix is epoxy. The number of plies of the specimen is 80 and the average thickness of each ply is 0.125 mm. The material is assembled in different periodic stacking sequences of fiber directions (45°/0°) to form multilayer structures, as shown in Figure 5. According to the NDT report that was provided by the manufacturer, the porosity of the thick section of the CFRP specimen is nearly zero, which is tested by using an industrial ultrasonic immersion scanner. The purpose of choosing such a zero porosity specimen is to eliminate the effect of the original manufacturing defects in CFRP, so that the test results of the simulated defects can be more reliable.

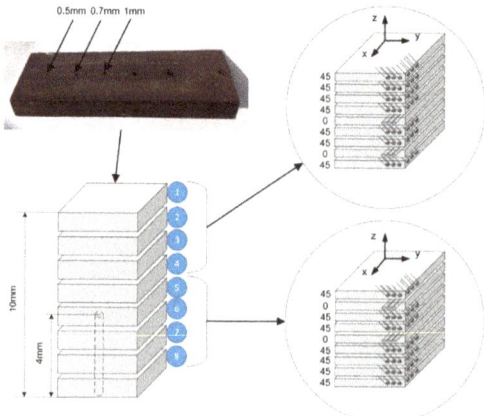

Figure 5. Lay-up of multidirectional $[45_4/0/45_2/0/45]_{4s}$ CFRP, total thickness 10 mm.

3.2. Test Method

The flat bottom holes of different diameters have been drilled to simulate discrete defects in CFRP, as shown in Figure 5. The depth is 4mm from the opposite surface of the ultrasonic probe for all

Appl. Sci. **2019**, *9*, 1183

artificial defects. Defect-free areas and areas with defects of different sizes have been detected multiple times by the ultrasonic pulse echo method, as shown in Table 1.

Table 1. Arrangement of experiments.

Experiment Set	Defect Diameter (mm)	Test Times
Defect-free-1	0	100
Defect-free-2	0	100
Defect-free-3	0	100
Defective-1	0.5	100
Defective-2	0.7	100
Defective-3	1	100

4. Results and Discussion

4.1. Recurrence Analysis of Defect-Free Areas

The purpose of this paper is to evaluate the discrete defects in thick multilayer CFRP by recurrence analysis of the second half of backscattered signals. The signal modality is affected by not only defects, but also material structures. Therefore, an area without artificial defects in the specimen has been analyzed first. Four signals have been randomly selected, and the second half of backscattered signals was magnified, as shown in Figure 6. The first half of each signal has the same resonance noise structure and large amplitude, while the second half has smaller amplitude and seem to be different from each other.

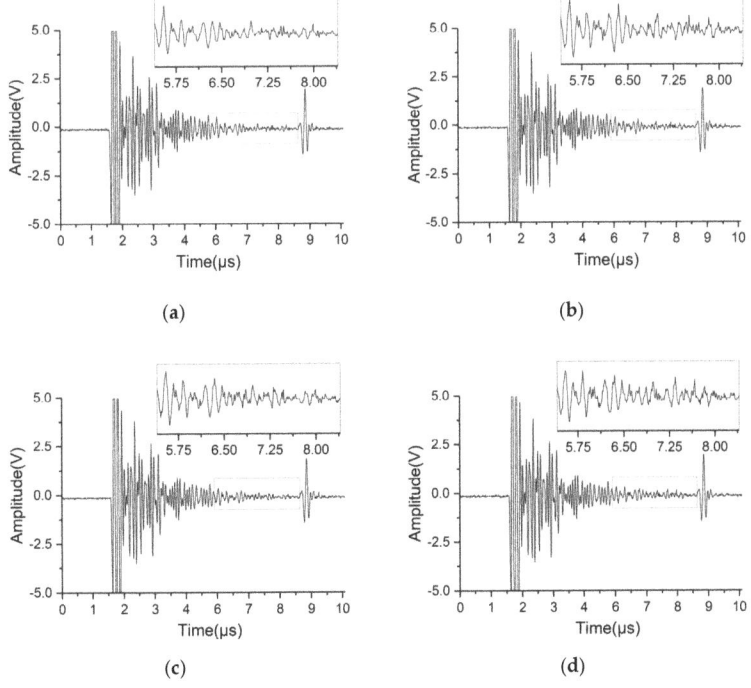

Figure 6. Ultrasonic signals randomly selected in results of experiment set Defect-free-1. (**a**) signal 1 (**b**) signal 2 (**c**) signal 3 (**d**) signal 4.

For further recurrence analysis of the signal, the mutual information method has been used for determination of the delay time (Lag). The relationship between the mutual information value and Lag has been obtained, as shown in the Figure 7a. The first local minimum of the mutual information value was the optimal time delay value. As a result, the optimal delay time $\tau = 3$.

Next, the false nearest neighbor algorithm has been used to obtain the embedding dimension. Figure 7b shows the relationship between the ratio of false nearest neighbors and the value of the embedded dimension. The embedding dimension is considered to be the best when the ratio of false nearest neighbors is close to zero. Thus, the optimal embedding dimension $m = 5$.

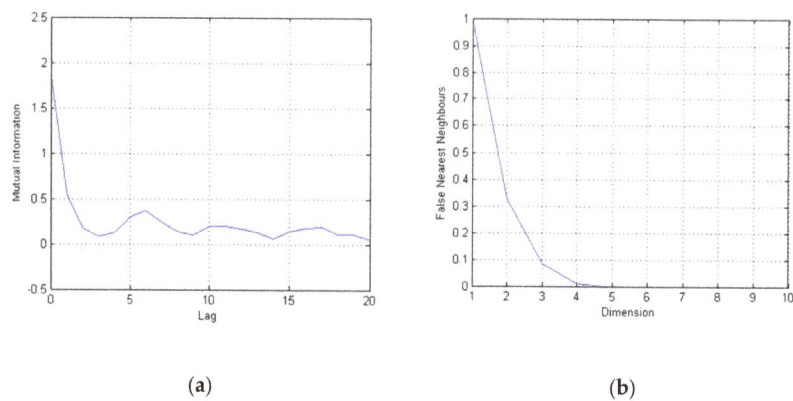

Figure 7. (a) The relationship between the mutual information value and the delay time (Lag). (b) The relationship between the ratio of false nearest neighbors and the value of embedded dimension.

It has been verified by calculation that the embedding dimension and the delay time of all signals are the same. Nonoptimal embedding parameters may cause many small blocks or even diagonal lines perpendicular to the main diagonal line [24], which should be carefully checked in the RP.

A lot of methods could be used to select the threshold, which need to be determined according to specific problems. The most commonly used threshold is that with a fixed value. In order to find the statistical behavior of the RQA parameters, while both the DET and LAM are variables that are based on the total amount of recurrence points according to Equations (4) and (5), the recurrence point rate RR needs to be determined first. Usually, RR takes 0.1; the threshold should be adjusted so that the statistical mean of RRs of all signals is 0.1. The results show that the requirement can be met when the threshold is 0.28 for defect-free signals.

The RPs of the second half of the four signals were calculated using the parameters above, as shown in Figure 8. It can be seen that the structures of the figures are similar and mainly composed of some special structures: short diagonal lines, small black blocks, and vertical and horizontal lines. According to the meaning of structures in RP, short diagonal lines and small blocks can be regarded as the suggestion of chaos. Short diagonal lines indicate that the periodic behavior of the ultrasonic pulse only lasts for a short time, and small black blocks represent different states of the ultrasonic pulse. Ultrasonic pulses travel the material thickness from the front surface echo to the back wall echo twice and may reflect multiple times in the material. Thus, it is reasonable to the existence of chaotic components in the ultrasonic signal, especially those with large number of plies and complex lay-up. Besides, vertical or horizontal lines represent time segments that remain unchanged or change very slowly, and they are typical behaviors of the state of the laminate. This is consistent with the multilayer structure of the composite, but the vertical lines do not exactly correspond to the lay-up of the material. One possible reason is related to the wavelength of the ultrasonic pulse. For the frequency 7.5 MHz, the wavelength of the ultrasonic pulse is approximately 0.4 mm, which is larger than the thickness of the fiber layer 0.1 mm. Thus, some detailed vertical structures may be missing. Although not able to get all of the details of the material structure, recurrence analysis can reveal the modality of the signal and it is sufficient for the detection of discrete defects.

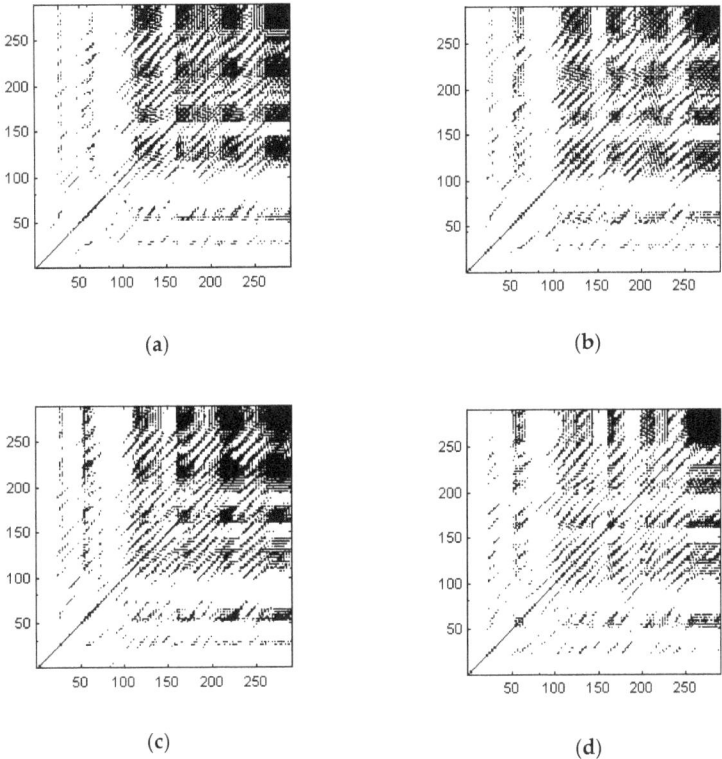

Figure 8. Recurrence plots (RPs) of the second half of backscattered signals of Defect-free-1. (**a**) signal 1 (**b**) signal 2 (**c**) signal 3 (**d**) signal 4.

The density of recurrence points indicates the structure of recurrence matrix and it may be a representation of instability of the signal. Analyze the first and the second half of all 100 Defect-free-1 signals separately. The RRs were calculated while using the same embedding dimension and delay used in RPs. The threshold was chosen so that the mean value of RRs of all 100 results equal 0.1, and the statistical results are shown in Figure 9. The median (50%) is a reflection of the concentration trend. Medians of both parts of the backscattered signal are approximately equal to 0.1, the mean of all RRs. The interquartile range (IQR, 75–25%) indicates the dispersion of variables in the statistics, and it can be a good representation of the robustness of the signal. The IQR of the first half is 0.0093, which is far less than the value of the second half 0.0468. A large IQR indicates that the instability of the second half of the backscattered signal might be due to multiple reflections of the ultrasonic pulse in the multilayered structure.

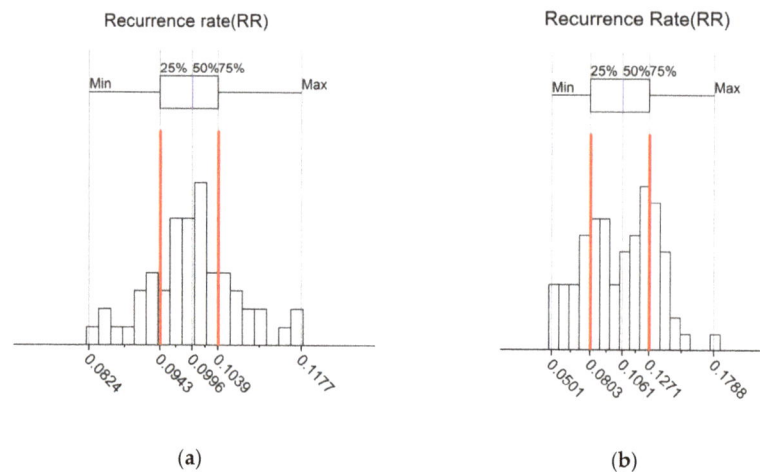

Figure 9. Box charts of recurrence rate (RR) of 100 Defect-free-1 signals: (**a**) the first half and (**b**) the second half.

Apply the method onto signals that were acquired in experiment sets Defect-free-1, Defect-free-2, and Defect-free-3. The RQA variables of the second half of backscattered signals were calculated using the same embedding dimension and delay above, while the threshold was chosen so that the mean value of RRs of all results equal to 0.1. The statistics of RR, DET, and LAM were shown in Figure 10. The median and IQR of RQA variables of different defect-free areas remain basically unchanged, while the max and min values vary greatly. When considering that the selected defect-free parts of the specimen have the same structure, the median and IQR of RQA variables are proper for characterizing the behavior of ultrasonic pulses in composite.

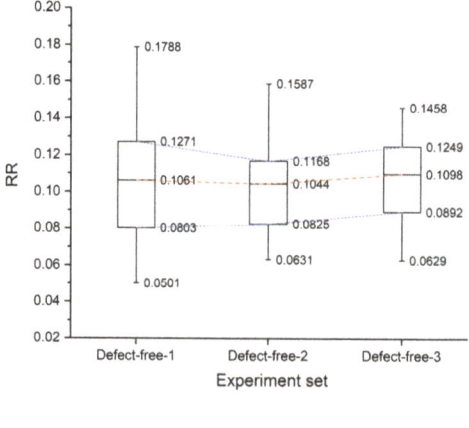

(**a**)

Figure 10. *Cont.*

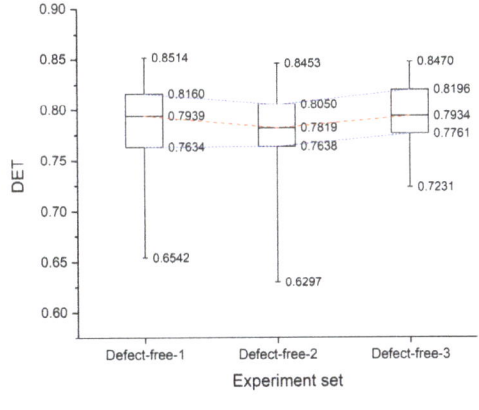

Figure 10. Statistical results of recurrence quantification analysis (RQA) variables of defect-free areas; (a) RR; (b) percent determinism (DET); and, (c) laminarity (LAM).

4.2. Recurrence Analysis of Defect Areas

According to the RPs of defect-free areas in Figure 8, chaotic behavior has been found in the ultrasonic signal. Despite the low porosity of the specimen, scattering and data noise may exist and form the feeble differences of cases in Figure 8, together with chaos. The statistical results of the RQA variables in Figure 10 reveal that the difference in RPs of different defect-free areas are in a relatively constant range, and the range can be described with the median and IQR of RQA variables. In other words, each value in the range of RQA variables represents a state of the system and the defect-free ultrasonic pulse echo system can be described by a set of RQA statistics: $RR_{median} = 0.1$, $RR_{IQR} \approx 0.4$; $DET_{median} = 0.79$, $DET_{IQR} \approx 0.45$; $LAM_{median} = 0.85$, and $LAM_{IQR} \approx 0.4$.

In order to discover the effect of defects on the states of the ultrasonic pulse echo system, the areas where the defects of 0.5 mm, 0.7 mm, and 1 mm were located were tested and the time series are shown in Figure 11. The signals of 0mm were from experiment set Defect-free-1, while those of 0.5 mm from Defective-1, 0.7 mm from Defective-2, and 1 mm from Defective-3. No recognizable defect echo could be found in the waveform either. The RPs of the second half of each signal were calculated and the results are shown in Figure 12. The plots have similar structures and they are mainly composed of diagonal lines and white bands. It can be inferred that the reflections of ultrasonic pulse caused by discrete defects in the thick CFRP do not appear to be local echo with a high amplitude in the time domain waveform. The effect of the defect is not the form of changing the structure of the signal.

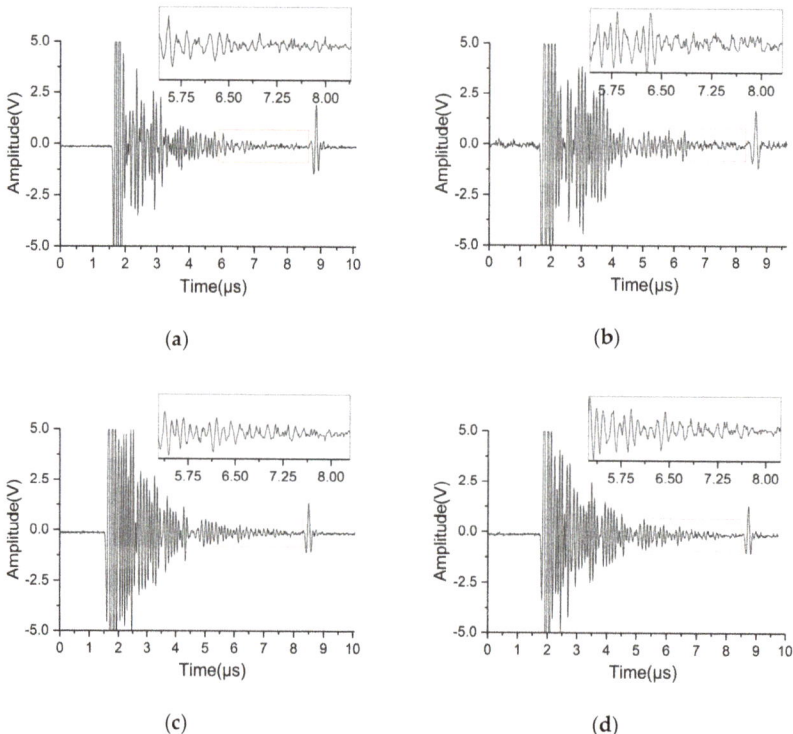

Figure 11. Ultrasonic signals of defect areas with different diameter: (**a**) 0 mm; (**b**) 0.5 mm; (**c**) 0.7 mm; and, (**d**) 1 mm.

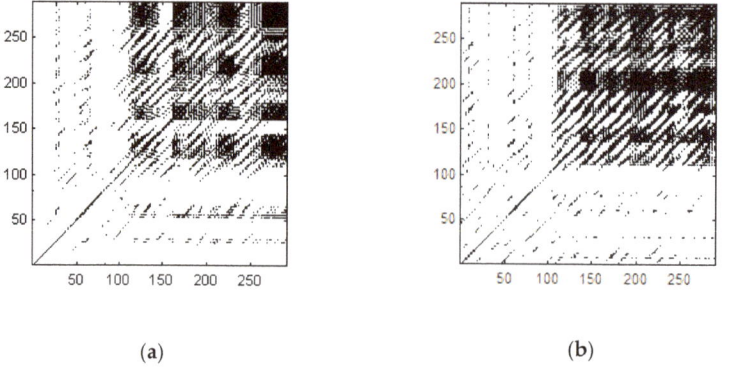

Figure 12. *Cont.*

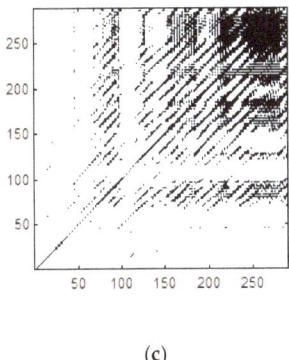
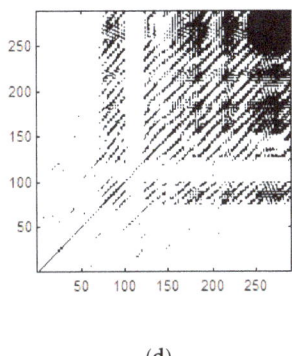

(c)   (d)

Figure 12. RPs of the second half of backscattered signals of defect areas with different diameter: (**a**) 0 mm; (**b**) 0.5 mm; (**c**) 0.7 mm; and, (**d**) 1 mm.

Calculate the RQA variables of the second half of the defect areas with the same embedding dimension and delay in 4.1, and the threshold was chosen so that the mean value of RRs of all results equal to 0.1. The statistics of RR, DET, and LAM are shown in Figure 13. In Figure 13a, the RR_{median} of different defect diameter are approximately equal, while the RR_{IQR} decreases as the defect size increases and tend to be constant to 0.01, which is much smaller than the defect-free RR_{IQR} of 0.04. In Figure 13b, DET_{IQR} of different defect diameter are approximately equal while the DET_{median} decreases as the defect size increases. The decrease is rather small and the DET_{median} tend to be constant around 0.75 which is sufficiently different from the value of defect-free areas. The defect size here is 0.7 mm, which is about twice the wavelength of the ultrasonic pulse. The same trend can be seen in the statistics of LAM in Figure 13c.

When comparing with results of defect-free areas in Figure 10, the statistical values of the RQA variables of defective areas are different and related to the size of the defect. For defect-free areas, the second half of backscattered signals is rather instable and shows chaotic characteristics. As the defect size increases, no significant change in chaotic characteristics is found, while the instability of the signal due to scattering and data noise is weakening. That is, the effect of defects on the signal modality is like a stabilizer, which makes the chaotic structure of the signal clearer. The larger the defect size, the more obvious the stabilization effect until the defect size is about twice the wavelength of the ultrasonic pulse, and in this paper, defects of 0.7 mm and 1 mm can be distinguished in the results of RQA, while 0.5 mm cannot.

In summary, using statistics of RQA variables, it is able to characterize the instability of the signal and reveal the effect of discrete geometry variation in the form of blind holes. The chaotic structure of the signal is more stable due to the presence of simulated defects and it is related to the size of the defect.

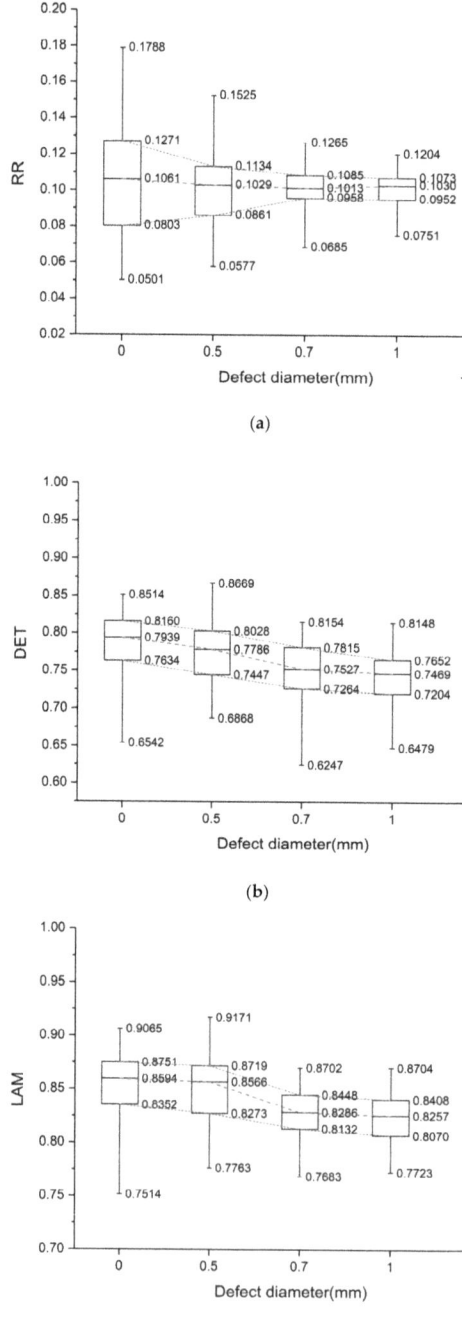

Figure 13. Statistical results of RQA variables of defect and defect-free areas: (**a**) RR; (**b**) DET; and, (**c**) LAM.

5. Conclusions

In this paper, a nonlinear method has been proposed for charactering discrete defects in thick multilayer composites. The second half of the backscattered signal is rather irregular, while the RP is able to reveal the unstable chaotic behavior of ultrasonic pulse and the median and IQR of RQA variables can be used for the description of system states. The method shows the possibility to detect blind holes of small diameter from ultrasonic pulsed-echo inspections. The maximum identifiable defect size may be related to the center frequency of the probe, which needs to be proved by future experiments of different probes and sizes of defects. The results of the nonlinear method and RQA variables can be used as a reference to detect the finite holes of minimum diameter 1 mm at depth of 5–10 mm in thick composites

It is necessary to detect the defect-free areas in the specimen under the uniform test conditions each time the method is used and the previously obtained defect-free data cannot be used as a standard, since the existence of the defect is discriminated with RQA statistics in comparison to those of the defect-free region. Besides, there is a gap between the blind holds and the real discrete defects. The proposed method is not currently capable of detecting real internal defects that are typical of composites with a considerable level of confidence. An artificial specimen with defects made with embedded thin defects or small thin entrapped voids can be produced in further work for the detection of larger voids or small delamination using recurrence analysis.

Author Contributions: Conceptualization, G.T., X.Z. (Xiaojun Zhou) and C.Y.; Formal analysis, G.T.; Funding acquisition, C.Y.; Investigation, G.T. and C.Y.; Methodology, G.T. and X.Z. (Xiang Zeng); Resources, X.Z. (Xiaojun Zhou) and C.Y.; Supervision, X.Z. (Xiaojun Zhou); Writing—original draft, G.T.; Writing—review & editing, G.T., X.Z. (Xiaojun Zhou), C.Y. and X.Z (Xiang Zeng).

Funding: This research was funded by 1: The Fundamental Research Funds for the Central Universities, No.2018QNA4001; 2: Zhejiang Provincial Natural Science Foundation of China under Grant No.LY18E050002.

Acknowledgments: The programs used in this paper are based on the CRP Toolbox for matlab.

Conflicts of Interest: The authors declare no conflict of interest.

References

1. Cawley, P.; Adams, R.D. Defect types and non-destructive testing techniques for composites and bonded joints. *Mater. Sci. Technol.* **1989**, *5*, 413–425. [CrossRef]
2. Scott, I.G.; Scala, C.M. A review of non-destructive testing of composite materials. *NDT Int.* **1982**, *15*, 75–86. [CrossRef]
3. Ibrahim, M.E. Nondestructive evaluation of thick-section composites and sandwich structures: A review. *Compos. Part A* **2014**, *64*, 36–48. [CrossRef]
4. Li, C.; Pain, D.; Wilcox, P.D. Imaging composite material using ultrasonic arrays. *NDT E Int.* **2013**, *53*, 8–17. [CrossRef]
5. Ibrahim, M.E.; Smith, R.A.; Wang, C.H. Ultrasonic detection and sizing of compressed cracks in glass-and carbon-fibre reinforced plastic composites. *NDT E Int.* **2017**, *92*, 111–121. [CrossRef]
6. Smith, R.A.; Nelson, L.J.; Mienczakowski, M.J. Automated analysis and advanced defect characterisation from ultrasonic scans of composites. *Insight Non-Destruct. Test. Cond. Monit.* **2009**, *51*, 82–87. [CrossRef]
7. Wang, L.; Rokhlin, S.I. Ultrasonic wave interaction with multidirectional composites: Modeling and experiment. *J. Acoust. Soc. Am.* **2003**, *114*, 2582. [CrossRef] [PubMed]
8. Dominguez, N.; Mascarot, B. Ultrasonic Non-destructive inspection of localized porosity in composite materials. In Proceedings of the Ninth European Conference on Non-Destructive Testing (ECNDT), Berlin, Germany, 25–29 September 2006.
9. Demirli, R.; Saniie, J. Model-based estimation of ultrasonic echoes. Part I: Analysis and algorithms. *IEEE Trans. Ultrason. Ferroelectr. Freq. Control* **2001**, *48*, 787–802. [CrossRef] [PubMed]
10. Demirli, R.; Saniie, J. Model-based estimation of ultrasonic echoes. Part II: Nondestructive evaluation applications. *IEEE Trans. Ultrason. Ferroelectr. Freq. Control* **2001**, *48*, 803–811. [CrossRef] [PubMed]

11. Hagglund, F.; Martinsson, J.; Carlson, J.E. Model-based estimation of thin multi-layered media using ultrasonic measurements. *IEEE Trans. Ultrason. Ferroelectr. Freq. Control* **2009**, *56*, 1689–1702. [CrossRef] [PubMed]
12. Marwan, N.; Riley, M.; Giuliani, A. *Translational Recurrences: From Mathematical Theory to Real-World Applications*; Springer Publishing Company Incorporated: New York, NY, USA, 2014.
13. Webber, C.; Marwan, N. *Recurrence Quantification Analysis-Theory and Best Practices*; Springer: Berlin, Germany, 2015. [CrossRef]
14. Carrión, A.; Genovés, V.; Gosálbez, J.; Miralles, R.; Payá, J. Ultrasonic signal modality: A novel approach for concrete damage evaluation. *Cem. Concr. Res.* **2017**, *101*, 25–32. [CrossRef]
15. Carrión, A.; Miralles, R.; Lara, G. Measuring predictability in ultrasonic signals: An application to scattering material characterization. *Ultrasonics* **2014**, *54*, 1904–1911. [CrossRef] [PubMed]
16. Brandt, C.; Maaß, P. Recurrence Quantification Analysis for Non-Destructive Evaluation with an Application in Aeronautic Industry. In Proceedings of the 19th World Conference on Non-Destructive Testing, Munich, Germany, 13–17 June 2016.
17. Brandt, C. Recurrence quantification analysis as an approach for ultrasonic testing of porous carbon fibre reinforced polymers. In *Recurrence Plots and Their Quantifications: Expanding Horizons*; Springer International Publishing: Cham, Switzerland, 2016. [CrossRef]
18. Kantz, H.; Schreiber, T. *Nonlinear Time Series Analysis*; Cambridge University Press: Cambridge, UK, 2003. [CrossRef]
19. Takens, F. Detecting strange attractors in turbulence. *Lect. Notes Math.* **1981**, *898*, 366–381. [CrossRef]
20. Marwan, N.; Romano, M.C.; Thiel, M. Recurrence plots for the analysis of complex systems. *Phys. Rep.* **2007**, *438*, 237–329. [CrossRef]
21. Eckmann, J.P.; Kamphorst, S.O.; Ruelle, D. Recurrence plots of dynamical systems. *Europhys. Lett.* **2007**, *4*, 973–977. [CrossRef]
22. Hobbs, B.; Ord, A. Nonlinear dynamical analysis of GNSS data: Quantification, precursors and synchronisation. *Prog. Earth Planet. Sci.* **2018**, *5*, 36. [CrossRef]
23. Zbilut, J.P.; Thomasson, N.; Webber, C.L. Recurrence quantification analysis as a tool for nonlinear exploration of nonstationary cardiac signals. *Med. Eng. Phys.* **2002**, *24*, 53–60. [CrossRef]
24. Marwan, N. How to avoid potential pitfalls in recurrence plot based data analysis. *Int. J. Bifurc. Chaos* **2011**, *21*, 1003–1017. [CrossRef]

© 2019 by the authors. Licensee MDPI, Basel, Switzerland. This article is an open access article distributed under the terms and conditions of the Creative Commons Attribution (CC BY) license (http://creativecommons.org/licenses/by/4.0/).

Article

Nondestructive Ultrasonic Inspection of Composite Materials: A Comparative Advantage of Phased Array Ultrasonic

Hossein Taheri [1,*] and Ahmed Arabi Hassen [2,3]

1. Department of Manufacturing Engineering, Georgia Southern University, Statesboro, GA 30460, USA
2. Manufacturing Demonstration Facility (MDF), Oak Ridge National Laboratory (ORNL), Knoxville, TN 37932, USA; hassenaa@ornl.gov
3. Department of Mechanical, Aerospace and Biomedical Engineering, University of Tennessee, Estabrook Rd, Knoxville, TN 37916, USA
* Correspondence: htaheri@georgiasouthern.edu; Tel.: +1-912-478-7463

Received: 17 March 2019; Accepted: 16 April 2019; Published: 19 April 2019

Featured Application: The featured application of the proposed study is to develop the application and describe the advantages of phased array ultrasonic technique for the inspection of composite materials. The proposed method not only enhances the probability of detection of the defects in composite materials, but also increases the distance over which the defects are detectable with a single inspection location.

Abstract: Carbon- and glass fiber-reinforced polymer (CFRP and GFRP) composite materials have been used in many industries such as aerospace and automobile because of their outstanding strength-to-weight ratio and corrosion resistance. The quality of these materials is important for safe operation. Nondestructive testing (NDT) techniques are an effective way to inspect these composites. While ultrasonic NDT has previously been used for inspection of composites, conventional ultrasonic NDT, using single element transducers, has limitations such as high attenuation and low signal-to-noise ratio (SNR). Using phased array ultrasonic testing (PAUT) techniques, signals can be generated at desired distances and angles. These capabilities provide promising results for composites where the anisotropic structure makes signal evaluation challenging. Defect detection in composites based on bulk and guided waves are studied. The capability of the PAUT and its sensitivity to flaws were evaluated by comparing the signal characteristics to the conventional method. The results show that flaw sizes as small as 0.8 mm with penetration depth up to 25 mm can be detected using PAUT, and the result signals have better characteristics than the conventional ultrasonic technique. In addition, it has been shown that guided wave generated by PAUT also has outstanding capability of flaw detection in composite materials.

Keywords: phased array ultrasonic; composites; signal sensitivity; defect detection; nondestructive testing (NDT)

---

1. Introduction

Carbon fiber-reinforced polymer (CFRP) and glass fiber-reinforced polymer (GFRP) composite materials are widely used in a variety of applications such as aerospace structures, wind turbine blades, the automotive industry, and mass transit [1–4]. Nondestructive testing/evaluation (NDT/E) and inspection of these materials are necessary to control the quality of the parts and inspect for anomalies in the structures to prevent catastrophic failure. Nondestructive techniques are widely used for material evaluation and flaw detection [5,6]. Ultrasonic testing is one of the most commonly used NDT methods for various applications, where characteristics of ultrasonic signals, such as reflection and scattering of ultrasound waves, are used for material properties evaluation and flaw detection [5,7–10]. In ultrasonic testing, a piezoelectric transducer is commonly used for generation of compression or shear wave which are propagating through the inspected media. When these waves interact with media boundaries, they face reflection, transmission, and scattering from the boundaries [5]. These scattering characteristics, the speed of sound wave, and travelling time provide valuable information about the material properties and integrity. However, using conventional ultrasonic methods for composite inspection can be challenging due to the anisotropic nature of the composites structures [11–13]. Wave propagation in anisotropic composite structures is complex, and random scattering as well as high attenuation of ultrasonic waves reduce the probability of defect detection [14,15]. Several ultrasonic techniques have been used for inspection and characterization of composite materials. Castellano et. al. (2018) introduced a new experimental approach for the comparison between Quasi Static Indentation (QSI) damage and Low-Velocity Impact (LVI) damage in polymer composites starting from the results of ultrasonic goniometric immersion tests [16]. In their study, the differences and similarities between QSI and LVI damage starting from the analysis of the variations of the acoustic behavior and by using a suitable anisotropic damage model developed in the framework of the Continuum Damage Mechanics theory [16].

Phased array ultrasonic testing (PAUT) can overcome conventional ultrasonic method limitations by providing the capability of signal focusing and steering at desired angles and locations [17–19]. In PAUT, a series of ultrasonic elements in a phased array transducer can provide the option to activate each individual element in a programmed sequence [20,21]. A phased array unit includes a computer-based instrument capable of driving multielements, as well as receiving and digitizing the returning echoes based on the appropriate delay law for firing the elements. This is done by changing the time between the outgoing ultrasonic pulses of each element so that the superimposed wave front effectively steers and shapes the resultant final sound beam. This capability assists in generating the desired type of ultrasonic signal and improving the wave characteristics in comparison to the conventional single-element ultrasonic transducer. The PAUT method can also be used to generate guided waves [22–26]. Guided waves are another type of ultrasonic wave, which provide useable features for inspection of plate type structures. Guided waves can travel longer distances compared to the other types of ultrasonic waves and can cover more area of inspection, making faster inspections possible [19,27–29]. Chimenti (1997) comprehensively discussed the composite materials and their inspection and characterization using guided waves [30].

In this work, we first compare the defect detection capability and sensitivity of the PAUT signals with single element (conventional) ultrasonic (SEUT). The back wall reflection of bulk wave through the thickness of composite samples was used to study the signal characteristics of the PAUT and compare them with SEUT. The sensitivity of the signal to flaw detection was also studied using the response signal from the artificially made defects in composite parts. Next, guided wave modes were generated using the PAUT system for defect detection in sample plates. The guided waves generated using PAUT were used to show the feasibility of flaw detection on composite plates.

2. Materials and Sample Preparation

2.1. SEUT Versus PAUT Methods

GFRP plates, extracted from a wind turbine blade, were used for the experiments as shown in Figure 1a. The GFRP samples have various thicknesses of 4, 10, 12, 18, and 25 mm. In order to study the sensitivity of flaw detection in both the PAUT and SEUT methods, various sizes of holes were drilled on one side of the sample (with the largest thickness being 25 mm), as shown schematically in Figure 1b.

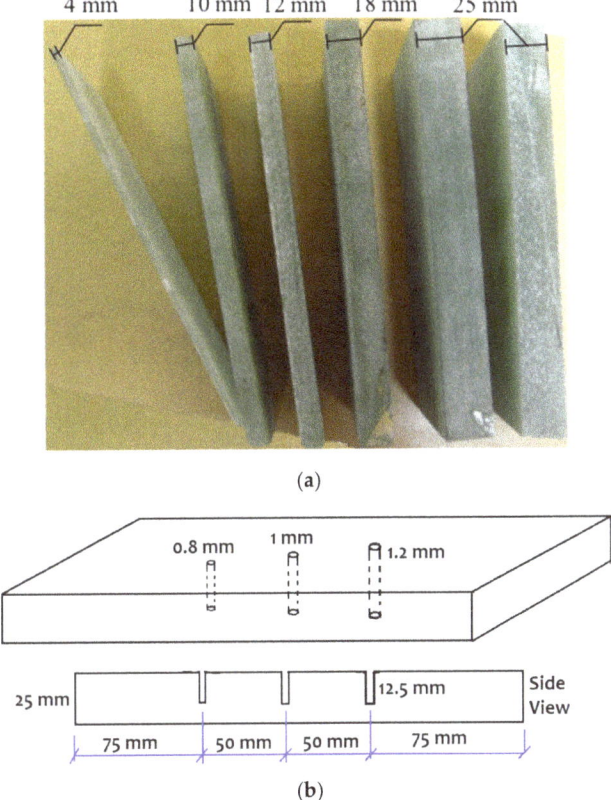

Figure 1. Samples used to evaluate capability and sensitivity of defect detection in single element ultrasonic (SEUT) versus phased array ultrasonic (PAUT) methods: (**a**) glass fiber reinforced polymer (GFRP) samples from a wind turbine blade and (**b**) schematic for the artificial hole locations in GFRP sample with 25mm thickness (Thk.). Sample size is 250 × 100 × 25 (L × W × Thk.) mm. Width of the sample is 100 mm and holes drilled in the middle of the width.

2.2. PAUT Guided Wave Method

Two different types of materials were used in this experiment, Aluminum (Al) and CFRP plates, as introduced in Table 1. The reason for selecting these materials for guided wave evaluation was that they were available in plate shape and desired thicknesses (in the range of 1 to 2 mm) for guided wave generation. For both Aluminum and CFRP samples, artificial defects, in the form of drilled holes, were made into the samples. Figure 2 shows a schematic for the location and depth of the artificial holes in both inspected samples.

Table 1. Test samples description used for the phased array ultrasonic (PAUT) guided wave method.

Sample Name	Material	Thickness (mm)
Al-1	Aluminum 6063 (Plate)	2.2
Al-2	Aluminum 6063 (Plate)	0.635
CFRP	Unidirectional Carbon Fiber Composite (5 layers of carbon fiber fabric)	1.0

3. Experimental Setup

3.1. SEUT Versus PAUT Methods

It is important to understand how far an ultrasonic signal can travel through the composite material while the back wall reflection is still detectable. This shows the capability of signal focusing and propagation for an ultrasonic setup. The SEUT experiments were performed using three different frequencies including 0.5, 1.0, and 1.5 MHz, where the attenuation of ultrasound signals at different frequencies was evaluated. In the PAUT experiments, a 1.5 MHz, 16-element transducer was used accompanying the related normal wedge. Both SEUT and PAUT transducer and setups are shown in Figure 3.

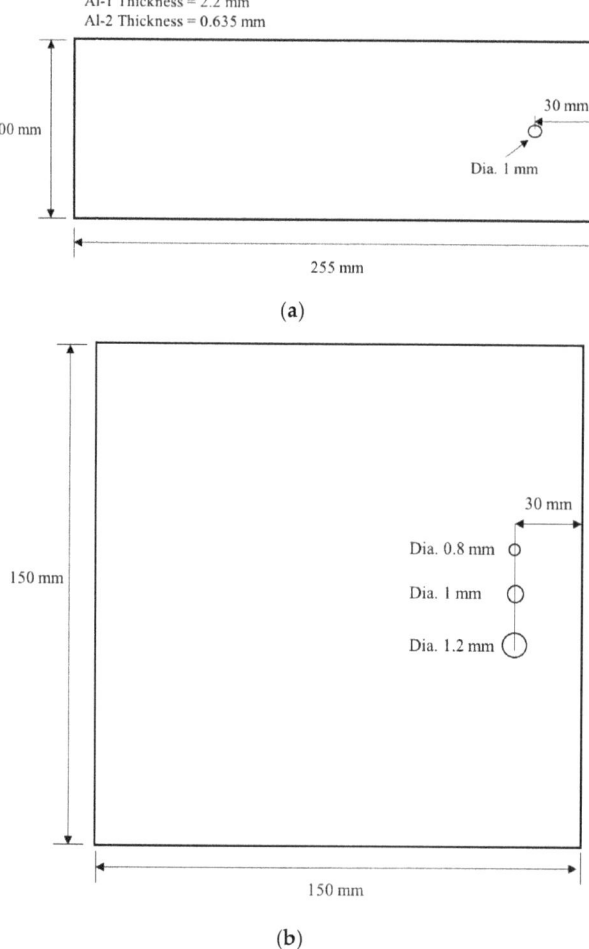

Figure 2. Schematic for the dimension and artificial holes locations in (**a**) Al-1 and Al-2 samples and (**b**) CFRP sample.

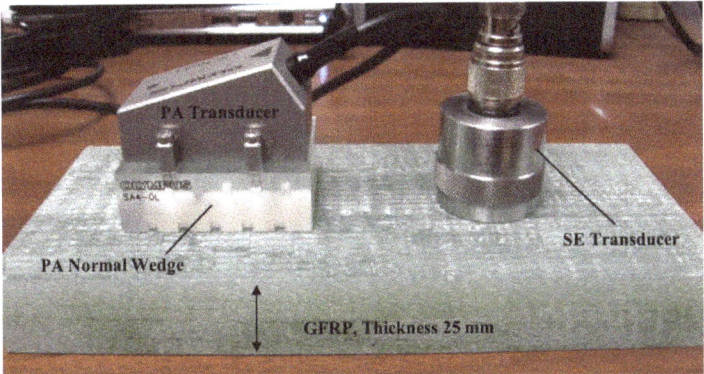

Figure 3. Experimental setup for capability and sensitivity evaluation: PAUT (left) and SEUT (right).

3.2. PAUT Guided Wave Method

Guided wave modes were generated on Al and CFRP plates by means of a commercially available phased array probe and wedges. The procedure of plate wave generation and parametric evaluation are described in detail in [8,18]. A 1.5 MHz phased array ultrasonic probe with 16 elements and related 60 degrees longitudinal wave wedge was used for guided wave generation and flaw detection. Figure 4 shows the setup used for inspecting the artificial defects (drilled holes) in CFRP sample. In CFRP sample, guided waves were generated in direction of the fibers.

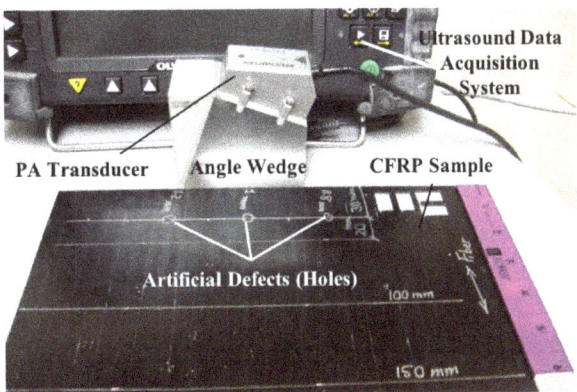

Figure 4. Experimental setup for flaw detection in CFRP sample using PAUT guided wave method.

4. Results and Discussions

4.1. SEUT Versus PAUT Methods

4.1.1. Focusing Depth Comparison

Table 2 shows the signal characteristics in terms of signal-to-noise ratio (SNR) for SEUT and PAUT. Data in Table 2 is plotted in Figure 5 and shows the relationship between the thicknesses of the GFRP plates (i.e., wave traveling distance) and travelling time of ultrasound wave. It can be observed that the velocities in the GFRP plate can be calculated as twice the slope of the graph, which are equal to $2 \times 1.57 = 3.15$ mm/µs for SEUT and $2 \times 1.59 = 3.18$ mm/µs for PAUT. Figure 6 shows an example for typical signals for the back wall reflection in SEUT and PAUT methods in the 12-mm-thick composite plate. As can be seen from the results in Table 2 and considering the form of the ultrasound signal shown in Figure 6, the features of the signals are clearer and better detectable using PAUT with lower gain values. The wave velocity is important information in determining the depth and location of the defects according to the ultrasound wave's traveling time.

Table 2. Signal characteristics of back wall reflection for SEUT method with different frequencies and PAUT method.

Frequency	Method	Sample	Thickness (mm)	Gain (dB)	Time (µs)	Signal-to-Noise Ratio
0.5 MHz	SEUT	GFRP	4	42	5.01	7.95
			10	55.8	9.87	3.99
			12	60.5	11.01	3.07
			18	61	14.73	2.98
			25	64.7	18.92	2.53
1 MHz	SEUT	GFRP	4	18.5	4.76	5.31
			10	36	8.56	11.93
			12	39.6	9.80	11.75
			18	41.2	13.60	11.10
			25	45.9	18.35	9.57
1.5 MHz	SEUT	GFRP	4	15.5	4.67	5.14
			10	37	8.48	11.75
			12	40.8	9.76	11.10
			18	43	13.55	9.51
			25	46.7	18.25	9.43
1.5 MHz	PAUT	GFRP	4	13	3.64	6.10
			10	27	7.56	5.44
			12	28.5	8.93	5.81
			18	36	12.49	3.39
			25	40	17.41	3.22

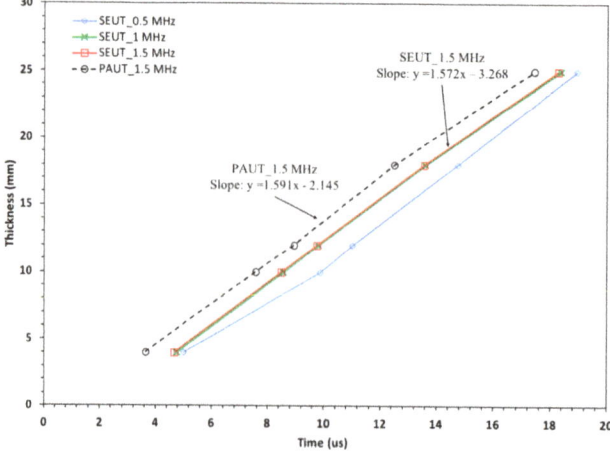

Figure 5. Back wall reflection experiment showing ultrasonic wave velocity evaluation measured by SEUT and PAUT methods.

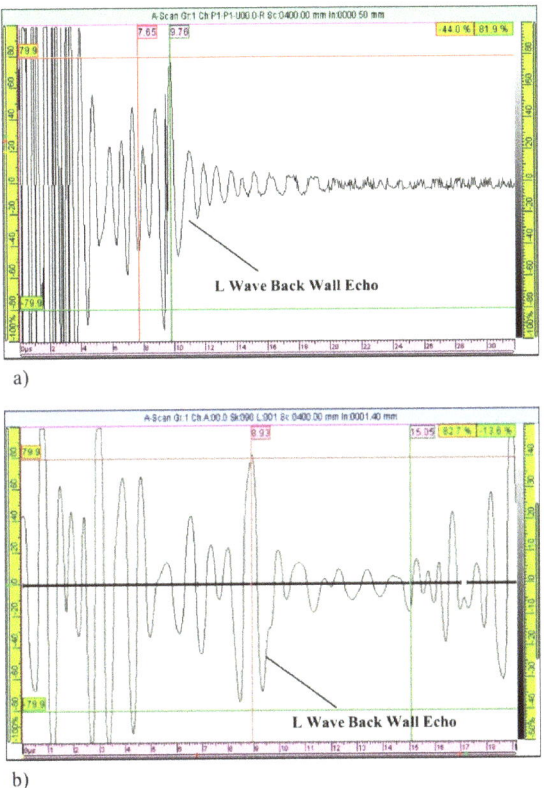

Figure 6. Back wall reflection signals for GFRP: (**a**) SEUT method and (**b**) PAUT method. (Freq. = 1.5 MHz, Thickness = 12 mm).

All the measured velocities are very close in value; however, the attenuation (i.e., gain values) is improved for PAUT when compared to SEUT at 1.5 MHz. On the other hand, the quantitative values in Table 2 show that SNR is, on average, two times larger for SEUT when comparing the peak of reflected signal to the background noise. However, it should be mentioned that the resolution of the peak and its location is much lower in SEUT which caused inaccuracy for detection purposes. Higher local value of SNR in SEUT can be attributed to the interference of the signals for each element in PAUT. Qualitatively, PAUT has more uniform and detectable signal with less jitter, specifically at larger thicknesses. Figure 7 shows the gain values (for different inspection frequencies) in order to reach detectable signal in different sample thicknesses. The plot shows that the SNR and signal's attenuation were improved in PAUT technique when compared to SEUT such that 7–20% less gain in value was required to have detectable signal in case of 1.5 MHz transducers.

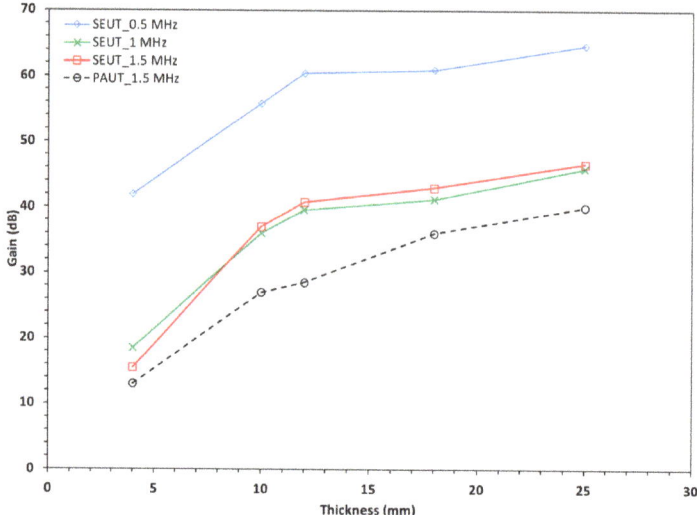

Figure 7. Gain (dB) value for back wall reflection detection in different sample thickness for attenuation characteristics evaluation using SEUT and PAUT techniques.

4.1.2. Sensitivity Comparison and Defect Detection

Figure 8 shows the signals associated with the defect (artificially drilled holes) reflections by SEUT and PAUT methods. The depth of the hole can be determined based on the obtained velocity values, as in Equations (1) and (2). Both SEUT and PAUT techniques provided results that are very close to what was obtained by real time x-ray imaging (i.e., 11.175 mm) for validation. It was observed that both SEUT and PAUT techniques can detect a 0.8 mm diameter hole as the minimum size and sensitivity limit; however, PAUT method provides approximately 15% higher SNR for the defect signal. We believe that in PAUT, lower SNR and better signal characteristics, such as higher focusing energy, could assist in detecting smaller-sized defect sizes, and this needs further experimental evaluation.

$$\text{Depth(SEUT)} = \frac{\text{Time} \times \text{Velocity}}{2} = \frac{7.12 \times 3.07}{2} = 10.9 \text{ mm} \qquad (1)$$

$$\text{Depth(PAUT)} = \frac{\text{Time} \times \text{Velocity}}{2} = \frac{7.05 \times 3.23}{2} = 11.4 \text{ mm} \qquad (2)$$

In Figure 8, PAUT has a clearer and more easily detectable reflection from the defect (reflector), as well as a better detectable back wall reflection. However, when looking at SEUT signal, due to less smoothness in signal from one transducer element, it is more difficult to identify these reflection locations. In addition, as can be seen from the data in Table 2, the gain value is a very important factor. However in some cases the SNR in SEUT looks to be higher, but it was obtained with higher gain value. This happened when a small decrease in the gain value, less than the values in Table 2, did not provide a good detectable signal.

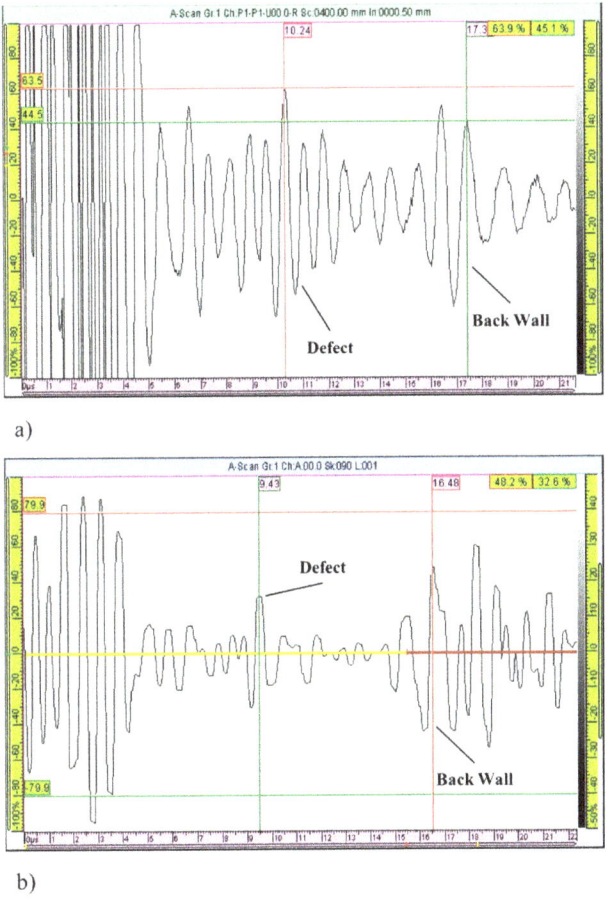

Figure 8. Response signal for GFRP: (**a**) SEUT method and (**b**) PAUT method. (Freq. = 1.5 MHz, Thickness = 25 mm, Hole Diameter = 0.8 mm, Hole Depth = 12mm).

4.2. PAUT Guided Wave Method

Tables 3 and 4 show the results for the signal response parameters for Al-1 and Al-2 samples. In the dispersion curve of guided wave modes in plates, the smaller values of "*fd*" (i.e., frequency × plate thickness) are more distinctive and, consequently, have a higher probability of detection [31]. In practical application specifically when the thickness of the plate structure is a fixed and known value, only the frequency of inspection can be changed. So, for thicker structures, one should use much lower frequencies, while for thinner structures, the range of possible frequencies will be wider and higher frequencies can be used to increase the resolution. The effect of "*fd*" value in response signals is presented in Tables 3 and 4. We find that, in lower "*fd*" values, the distance from which the signal from the defect is still detectable is longer. The phase velocity for the generated guided wave were calculated based on the theory and properties of the angle wedge. Based on these values, S0, A1, and S1 modes were possible for the Al-1 sample (*fd* = 3.3 MHz.mm), and A0 and S0 modes are possible for Al-2 sample (*fd* = 0.96 MHz.mm). The strongest reflection which also has the closest phase velocity value to the theory were identified as the dominate wave modes. In this case it was A1 for Al-1 sample, S0 for Al-2 sample, and S0 for CFRP sample. In Tables 3 and 4, signal parameters from the reflection of the edge of the plate close to the hole, and from the hole are presented. These signal parameters include the arrival time and the amplitude of the signal at the edge of the plate and the defect. Figure 9 shows the change of the signal's amplitude over the distance of the PAUT transducer from the edge for Al-1 and Al-2. For Al-1 it was noticed that the hole's signal has larger amplitude at a longer distance compared to Al-2. This is attributed to the interference of the stationary wedge reflection signal and the signal from the hole. Figure 10 shows typical signals for the experiments in Tables 3 and 4 for Al-1 and Al-2.

Table 3. Signal parameters for flaw detection in Al-1 sample.

Experimental Setup Parameters for PAUT Guided Wave Inspection of Al-1 Sample					
Frequency (MHz)	Thickness (mm)	Gain (dB)	Element Qty.[1]	fd^2	Element Step[3]
1.5	2.2	30	4	3.3	1

[1] Number of active elements at each sequence in phased array ultrasound transducer
[2] frequency × plate thickness (MHz.mm)
[3] Incremental steps in terms of number of elements at each sequence

Defect Detection Signal Characteristics								
Experimental Trials #	Hole Diameter (mm)	Signal	Distance of Transducer from The Edge (mm)	Arrival Time (us)		D_{Time} (us)	Amplitude (%)	
				Edge	Hole		Edge	Hole
1	1	Edge/Hole	50	58.07	47.61	10.5	53.9	46.6
2		Edge/Hole	100	87.99	70.27	17.7	33.1	24.3
3		Edge/Hole	150	109.77	92.34	17.4	10.3	21.1

Table 4. Signal parameters for flaw detection for Al-2.

Experimental Setup Parameters for PAUT Guided Wave Inspection of Al-2 Sample					
Frequency (MHz)	Thickness (mm)	Gain (dB)	Element Qty.[1]	fd^2	Element Step[3]
1.5	0.635	30	4	0.96	1

[1] Number of active elements at each sequence in phased array ultrasound transducer
[2] frequency × plate thickness (MHz.mm)
[3] Incremental steps in terms of number of elements at each sequence

Defect Detection Signal Characteristics								
Experimental Trial #	Hole Diameter(mm)	Signal	Distance of Transducer from The Edge (mm)	Arrival Time (us)		D_{Time} (us)	Amplitude (%)	
				Edge	Hole		Edge	Hole
1	1	Edge/Hole	50	39.19	28.15	11.0	100	15.5
2		Edge/Hole	100	57.78	46.45	11.3	100	11.5
3		Edge/Hole	175	85.08	74.34	10.7	82.7	7.8
4		Edge/Hole	200	94.09	83.05	11.0	77.7	4.5

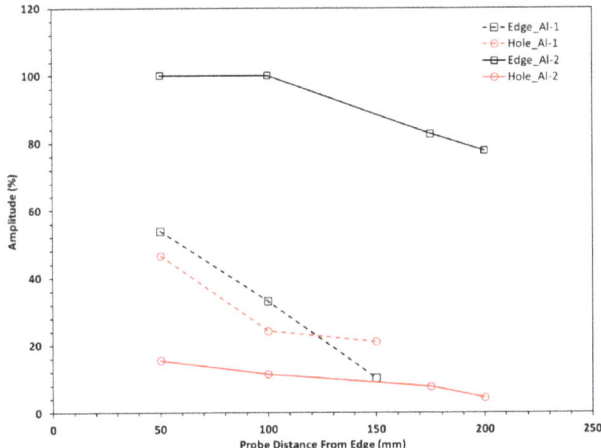

Figure 9. Change of signal amplitude over distance for Al-1 and Al-2 samples.

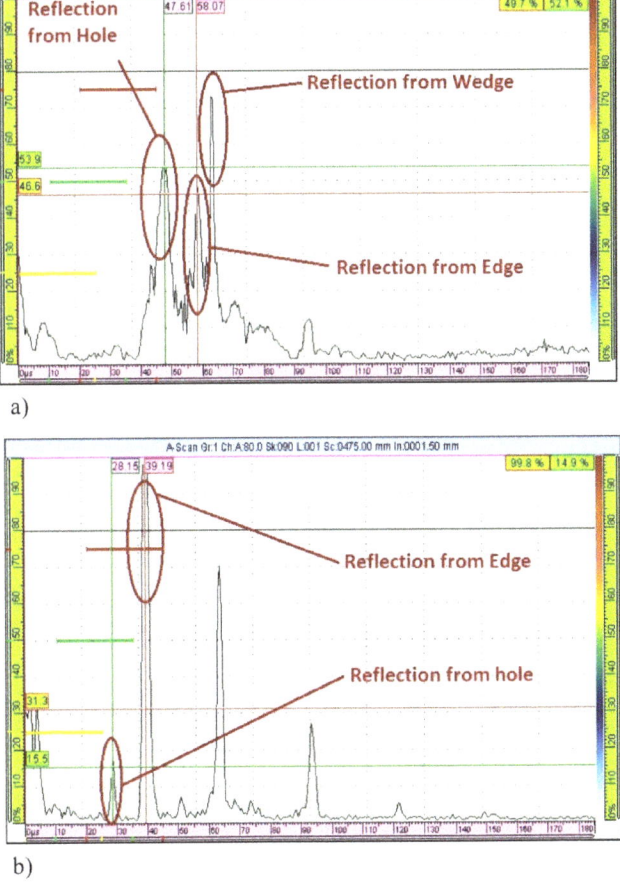

Figure 10. Typical PAUT guided wave signal of flaw detection for (**a**) Al-1 sample and (**b**) Al-2 sample.

Table 5 shows the results of PAUT guided wave signal parameters for the CFRP sample. In Table 5, the arrival time of the reflection signal from the plate edge and the defect (hole) are presented (See Figure 2 for reference). The difference between the arrival times from the plate edge and from the defect (hole) was calculated. Considering the wave velocity, which was experimentally determined in a previous work [8], the distance between the plate edge and defect (hole) was determined experimentally. Comparing the distance between the plate and the defect (hole) which was determined experimentally with the actual location (as designed = 30 mm) shows that the location of the defect (hole) can be determined using the proposed technique within an acceptable range. When the size of the defect (hole) is larger, there is a better probability of detection, and the accuracy in determining the location of the defect (hole) is higher. In addition, when there is less interference between wave modes, such as in the case of larger distances, there is higher accuracy of detection. Better accuracy at larger distances occurs because, when the travelling distance for the guided waves is short, these wave modes are not stabilized and have many overlaps and low signal-to-noise ratio which cause higher inaccuracy. Figure 11 shows a typical signal for the experimental result listed in Table 5 for CFRP. Figure 12 shows the change in signal amplitude at different defect (hole) sizes for the experiments in Table 5 for CFRP. The amplitude of the signal from the plate edge is inversely proportional to the hole diameter. However, the amplitude of the signal from the hole is directly proportional to the hole diameter. As the hole diameter increases (i.e., larger defect), a larger part of the ultrasonic energy is reflected by the defect (hole), and consequently a smaller part will hit the edge.

Table 5. Signal parameters for flaw detection for CFRP.

Experimental Setup Parameters for PAUT Guided Wave Inspection of CFRP					
Frequency (MHz)	Thickness (mm)	Gain (dB)	Element Qty.[1]	fd^2	Element Step[3]
1.5	1	35	4	1.5	1

[1] Number of active elements at each sequence in phased array ultrasound transducer
[2] frequency × plate thickness (MHz.mm)
[3] Incremental steps in terms of number of elements at each sequence

Defect Detection Signal Characteristics							
Experimental Trial #	Hole Diameter(mm)	Signal	Distance of Transducer from The Edge(mm)	Arrival Time (us) (Edge/Hole)		D_{Time} (us)	D_{dist} (mm)
				Edge	Hole		
1	0.8	Edge/Hole	50	25.64	13.81	11.83	48.9
2		Edge/Hole	100	35.6	24.54	11.06	45.7
3		Edge/Hole	150	46.11	37.08	9.03	37.3
1	1	Edge/Hole	50	26.76	14.07	12.69	52.5
2		Edge/Hole	100	35.62	26.11	9.51	39.3
3		Edge/Hole	150	45.97	37.28	8.69	35.9
1	1.2	Edge/Hole	50	24.52	14.11	10.41	43.0
2		Edge/Hole	100	36.24	26.89	9.35	38.6
3		Edge/Hole	150	46.25	37.36	8.89	36.7

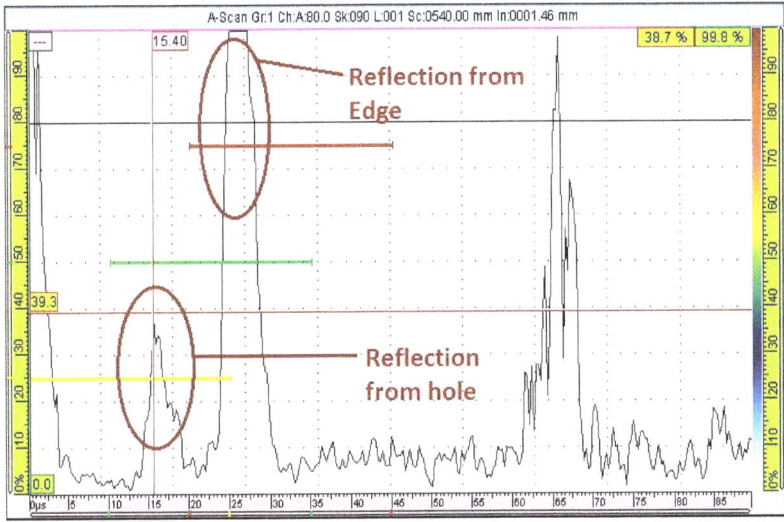

Figure 11. Typical PAUT guided wave signal of flaw detection for CFRP sample.

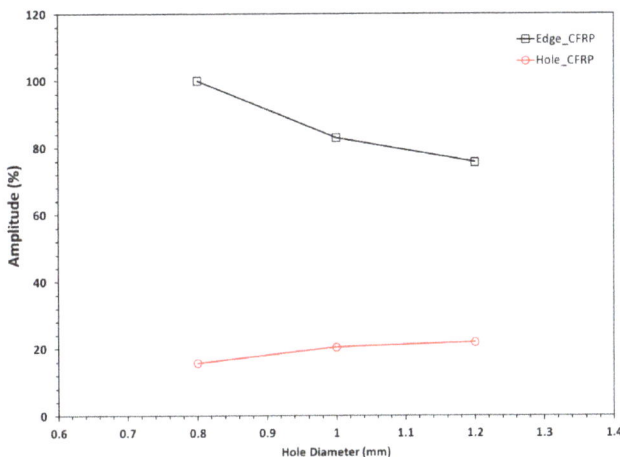

Figure 12. Change of signal amplitude over distance for CFRP samples.

5. Conclusions

Conventional (single-element) ultrasonic testing and phased array ultrasonic testing (PAUT) methods are evaluated for inspection of composite materials. The capability comparison tests for waves traveling through the composite materials indicate that thickness of up to 25 mm could be tested in both SEUT and PAUT methods; however, the stability of the signal parameters is higher in PAUT and detectable signal can be observed at lower gain values. The calculated velocity from the capability experimental part was 3.1 mm/μs, which is close to the estimated and expected velocities in composite plates and was used to identify the flaw's location. From the sensitivity comparison experimental results, it can be seen that a 0.8 mm diameter hole can be detected as the minimum size by both SEUT and PAUT, but PAUT generally has a better signal regarding SNR. However, PAUT does not increase the sensitivity by a big factor, but because of lower noise and jitter and better signal characteristics it may be possible to find smaller defect sizes such as 0.7 or 0.6 mm with PAUT as well.

Guided waves can also be generated using phased array ultrasonic probes and wedges with lower frequencies. Experimental results show that the different size of flaw (0.8, 1, and 1.2 mm diameter holes) can be detected by means of generated guided waves with the PAUT method. While the determination of the exact location of the flaw is affected by the dispersion characteristics of the guided waves, PAUT is a promising technique for detecting the size and location of defects in CFRP and GFRP composite materials.

Author Contributions: Conceptualization, H.T. and A.A.H.; Methodology, H.T.; Formal Analysis, H.T.; Investigation, H.T. and A.A.H.; Data Curation, H.T.; Writing—Original Draft Preparation, H.T.; Writing—Review and Editing, A.A.H.; Visualization, H.T. and A.A.H.; Supervision, A.A.H.

Funding: Research sponsored by the U.S. Department of Energy, Office of Energy Efficiency and Renewable Energy, Advanced Manufacturing Office, under contract DE-AC05-00OR22725 with UT-Battelle, LLC.

Acknowledgments: A debt of gratitude is owed to OLYMPUS NDT technical support and sales team for all their contributions to provide required equipment, support, and information.

Conflicts of Interest: The authors declare no conflict of interests. Notice of Copyright: This manuscript has been authored by UT-Battelle, LLC under Contract No. DE-AC05-00OR22725 with the U.S. Department of Energy. The United States Government retains and the publisher, by accepting the article for publication, acknowledges that the United States Government retains a non-exclusive, paid-up, irrevocable, world-wide license to publish or reproduce the published form of this manuscript, or allow others to do so, for United States Government purposes. The Department of Energy will provide public access to these results of federally sponsored research in accordance with the DOE Public Access Plan. (http://energy.gov/downloads/doe-public-access-plan).

References

1. Poudel, A.; Shrestha, S.S.; Sandhu, J.S.; Chu, T.P.; Pergantis, C.G. Comparison and Analysis of Acoustography with Other NDE Techniques for Foreign Object Inclusion Detection in Graphite Epoxy Composites. *Compos. Part B Eng.* **2015**, *78*, 86–94. [CrossRef]
2. Raišutis, R.; Jasiuniene, E.; Sliteris, R.; Vladišauskas, A. The review of non-destructive testing techniques suitable for inspection of the wind turbine blades. *Ultrasound* **2008**, *63*, 26–30.
3. Amenabar, I.; Mendikute, A.; López-Arraiza, A.; Lizaranzu, M.; Aurrekoetxea, J. Comparison and analysis of non-destructive testing techniques suitable for delamination inspection in wind turbine blades. *Compos. Part B Eng.* **2011**, *42*, 1298–1305. [CrossRef]
4. Adem, E.; Reddy, G.M.; Koricho, E.G.; Science, A.; Science, A.; Science, A.; Vehicle, C. Experimental Analysis of E-Glass/Epoxy & E-Glass/polyester Composites for Auto Body Panel. *Am. Int. J. Res. Sci. Technol. Eng. Math.* **2015**, *10*, 377–383.
5. Ensminger, D.; Bond, L.J. *Ultrasonics: Fundamentals, Technologies, and Applications*, 3rd ed.; CRC Press: Boca Raton, FL, USA, 2011.
6. Hübschen, G.; Altpeter, I.; Tschuncky, R.; Herrmann, H.-G. (Eds.) *Materials Characterization Using Nondestructive Evaluation (NDE) Methods*; Elsevier: Amsterdam, The Netherlands, 2016.
7. Taheri, H. Classification of Nondestructive Inspection Techniques with Principal Component Analysis (PCA) for Aerospace Application. In Proceedings of the ASNT 26th Research Symposium, Jacksonville, FL, USA, 13–16 March 2017; pp. 219–227.
8. Taheri, H. Utilization of Non-Destructive Testing (NDT) Methods for Composite Material Inspection (Phased array Ultrasonic). Master's Thesis, South Dakota State University, Brookings, South Dakota, 2014.
9. Wertz, J.; Homa, L.; Welter, J.; Sparkman, D.; Aldrin, J.C. Case Study of Model-Based Inversion of the Angle Beam Ultrasonic Response from Composite Impact Damage. *J. Nondestr. Eval. Diagn. Progn. Eng. Syst.* **2018**, *1*, 41001–41010. [CrossRef]
10. Taheri, H.; Delfanian, F.; Du, J. Ultrasonic phased array techniques for composite material evaluation. *J. Acoust. Soc. Am.* **2013**, *134*, 4013. [CrossRef]
11. Taheri, H.; Ladd, K.M.; Delfanian, F.; Du, J. Phased array ultrasonic technique parametric evaluation for composite materials. In *ASME International Mechanical Engineering Congress and Exposition, Proceedings (IMECE)*; American Society of Mechanical Engineers: New York, NY, USA, 2014; Volume 13, p. V013T16A028.

12. Caminero, M.A.; García-Moreno, I.; Rodríguez, G.P.; Chacón, J.M. Internal damage evaluation of composite structures using phased array ultrasonic technique: Impact damage assessment in CFRP and 3D printed reinforced composites. *Compos. Part B Eng.* **2019**, *165*, 131–142. [CrossRef]
13. Hassen, A.A.; Taheri, H.; Vaidya, U.K. Non-destructive investigation of thermoplastic reinforced composites. *Compos. Part B Eng.* **2016**, *97*, 244–254. [CrossRef]
14. Aldrin, J.C.; Wertz, J.N.; Welter, J.T.; Wallentine, S.; Lindgren, E.A.; Kramb, V.; Zainey, D. Fundamentals of angled-beam ultrasonic NDE for potential characterization of hidden regions of impact damage in composites. *AIP Conf. Proc.* **2018**, *1949*, 120005.
15. Toyama, N.; Ye, J.; Kokuyama, W.; Yashiro, S. Non-Contact Ultrasonic Inspection of Impact Damage in Composite Laminates by Visualization of Lamb wave Propagation. *Appl. Sci.* **2018**, *9*, 46. [CrossRef]
16. Castellano, A.; Fraddosio, A.; Piccioni, M.D. Quantitative analysis of QSI and LVI damage in GFRP unidirectional composite laminates by a new ultrasonic approach. *Compos. Part B Eng.* **2018**, *151*, 106–117. [CrossRef]
17. Taheri, H.; Delfanian, F.; Du, J. Acoustic Emission and Ultrasound Phased Array Technique for Composite Material Evaluation. In *ASME International Mechanical Engineering Congress and Exposition, Proceedings (IMECE): Advances in Aerodynamics*; American Society of Mechanical Engineers: New York, NY, USA, 2013; Volume 1, p. V001T01A015.
18. Bai, Z.; Chen, S.; Xiao, Q.; Jia, L.; Zhao, Y.; Zeng, Z. Compressive sensing of phased array ultrasonic signal in defect detection: Simulation study and experimental verification. *Struct. Health Monit.* **2017**, *17*, 434–449. [CrossRef]
19. Taheri, H.; Du, J.; Delfanian, F. Experimental Observation of Phased Array Guided Wave Application in Composite Materials. *Mater. Eval.* **2017**, *75*, 1308–1316.
20. Bolotina, I.; Borikov, V.; Ivanova, V.; Mertins, K.; Uchaikin, S. Application of phased antenna arrays for pipeline leak detection. *J. Pet. Sci. Eng.* **2018**, *161*, 497–505. [CrossRef]
21. Taheri, H.; Koester, L.; Bigelow, T.; Bond, L.J.; Braconnier, D.; Carcreff, E.; Dao, A.; Caulder, L.; Hassen, A.A. Fast Ultrasonic Imaging with Total Focusing Method (TFM) for Inspection of Additively Manufactured Polymer Composite Component. In Proceedings of the 27th ASNT Research Symposium, São Paulo, Brazil, 27–29 August 2018; pp. 212–220.
22. Drinkwater, B.W.; Wilcox, P.D. Ultrasonic arrays for non-destructive evaluation: A review. *NDT&E Int.* **2006**, *39*, 525–541.
23. Fromme, P.; Wilcox, P.D.; Lowe, M.J.S.; Cawley, P. On the development and testing of a guided ultrasonic wave array for structural integrity monitoring. *IEEE Trans. Ultrason. Ferroelectr. Freq. Control* **2006**, *53*, 777–785. [CrossRef] [PubMed]
24. Leleux, A.; Micheau, P.; Castaings, M. Long Range Detection of Defects in Composite Plates Using Lamb Waves Generated and Detected by Ultrasonic Phased Array Probes. *J. Nondestr. Eval.* **2013**, *32*, 200–214. [CrossRef]
25. Philtron, J.H.; Rose, J.L. Guided wave phased array sensor tuning for improved defect detection and characterization. In *Proc. SPIE 9063, Nondestructive Characterization for Composite Materials, Aerospace Engineering, Civil Infrastructure, and Homeland Security*; International Society for Optics and Photonics: San Diego, CA, USA, 2014; p. 906306.
26. Wang, W.; Zhang, H.; Lynch, J.P.; Cesnik, C.E.S.; Li, H. Experimental and numerical validation of guided wave phased arrays integrated within standard data acquisition systems for structural health monitoring. *Struct. Control Health Monit.* **2018**, *25*, e2171. [CrossRef]
27. Rose, J.L. Ultrasonic guided waves in structural health monitoring. *Key Eng. Mater.* **2004**, *270*, 14–21. [CrossRef]
28. Rose, J.L. *Ultrasonic Guided Waves in Solid Media*; Ultrasonic Guided Waves in Solid Media; Cambridge University Press: Cambridge, UK, 2014; Volume 9781107048959, pp. 1–512.
29. Wilcox, P.; Lowe, M.; Cawley, P. Effect of dispersion on long-range inspection using ultrasonic guided waves. *NDT E Int.* **2001**, *34*, 1–9. [CrossRef]
30. Chimenti, D.E. Guided waves in plates and their use in materials characterization. *Appl. Mech. Rev.* **1997**, *50*, 247–284. [CrossRef]

31. Rose, J.L. Successes and Challenges in Ultrasonic Guided Waves for NDT and SHM. In Proceedings of the National Seminar & Exhibition on Non-Destructive Evaluation, Pune, India, 4–6 December 2009.

© 2019 by the authors. Licensee MDPI, Basel, Switzerland. This article is an open access article distributed under the terms and conditions of the Creative Commons Attribution (CC BY) license (http://creativecommons.org/licenses/by/4.0/).

Article

Experimental Study of Defect Localization in a Cross-Ply Fiber Reinforced Composite with Diffuse Ultrasonic Waves

Qi Zhu [1,*], Yuxuan Ding [1], Dawei Tu [1], Haiyan Zhang [2] and Yue Peng [3,*]

1. School of Mechatronic & Automation Engineering, Shanghai University, Shanghai 200444, China; dyx2201@163.com (Y.D.); tdw@shu.edu.cn (D.T.)
2. School of Communication and Information Engineering, Shanghai University, Shanghai 200444, China; hyzh@shu.edu.cn
3. Logistics Engineering College, Shanghai Maritime University, Shanghai 201305, China
* Correspondence: Q_ZHU@shu.edu.cn (Q.Z.); yuepeng@shmtu.edu.cn (Y.P.); Tel.: +86-13166013707 (Q.Z.)

Received: 8 May 2019; Accepted: 3 June 2019; Published: 6 June 2019

Abstract: Diffuse wave inspection benefits from multiple scattering and is suitable for the nondestructive testing of complex structures with high sensitivity. This paper aims to localize the defect in a cross-ply carbon fiber reinforced polymer composite with the diffuse wave field experimentally based on the Locadiff technique. Firstly, the wave diffusivity and dissipation parameters are determined from the diffuse waveforms. Great dissipation is found for this composite plate due to its strong viscoelasticity, which makes the amplitude attenuate fast in a short propagation distance. The signal-to-noise ratios degrade significantly at off-axis directions so that only measurements along the X and Y axes are chosen. Secondly, the decorrelation coefficients are determined using the stretching technique. The decorrelation coefficients decrease initially due to the interaction between the wave fields and the defect and subsequently increase due to the low signal-to-noise ratio at the later time. Based on these data, a sensitivity time domain is chosen to center at $t = 50$ μs. Together with the defect sensitivity kernel calculated under constant diffusion property assumption, the defect is localized at [270 mm, 265 mm] compared to [300 mm, 280 mm] in the final reference state. This method is promising for early damage detection in fiber reinforced composite structures.

Keywords: diffuse ultrasonic waves; cross-ply fiber reinforced composite; defect localization

1. Introduction

Fiber reinforced polymer composites are becoming increasingly important in modern industries due to their high specific strength-weight ratio, anti-corrosion properties, and recyclability. They are manufactured using different processes such as resin transfer molding, extrusion, and automated tape placement, according to different property requirements. Recently, additive manufacturing, combined with various reinforcement control methods (e.g., standing-wave field [1], magnetic field [2], rotational deposition [3]), has come to be considered a cost-efficient method for composite design and manufacturing. During these processes, defects such as voids, cracks, and delaminations may be introduced into the part and lead to structure failure in service.

Defect localization and characterization are critical in modern lightweight structures made of fiber reinforced composites. The early detection of these defects is beneficial to the structural integrity and maintenance. Different methods have been exploited in recent years, including pulse-thermography [4], ultrasonic [5], X ray, acoustic emission [6,7], and electric resistance variation [8,9]. Among all these methods, ultrasonic testing is considered to be promising for in-situ or online inspection. Localization sensitivity can be improved through a higher ultrasonic frequency, but with regards heterogeneous materials such as concretes, biology tissues, and fiber reinforced composites, wave propagation becomes more complicated with increasing frequency. A full understanding of such process and proper signal processing methods is required to overcome or even benefit from scattering and attenuation phenomena.

The complex internal fiber distribution results in multiple scattering in the fiber reinforced composite. In such mediums, the whole waveform can be separated into the direct (ballistic, the first arrival) wave part and the diffuse wave part [10]. The direct wave is often strongly attenuated and only exists for a short distance that corresponds to the transport mean free path [11]. Subsequently, the energy of the direct waves rapidly converts into late-arriving diffuse waves [12]. The conventional ultrasonic inspection methods tend to lower the frequency to prevent the difficulties brought by attenuation and scattering. They often rely on the information from the direct wave part including the C scan method [13] and the Lamb wave inspection method [5]. The former has been widely accepted in aerospace industry to check defect existence in composite structures with considerable capital cost, and the latter is suitable for thin plate structures and can improve inspection efficiency. Nevertheless, careful interpretation of different propagation modes and boundary reflections/refractions are required [14]. By contrast, the diffuse wave is repeatable and independent of the direct wave path [15], and it can also be applied to structure inspection. Being treated as ambient noise [16] for a long time, the diffuse wave is found to contain valuable information, especially in seismology and civil engineering. Weaver and Lobkis [17] showed that the cross-correlation of two diffuse wave fields from the same excitation is equal to the direct response of one transducer to an impulse applied to the other. More recently, there has been a growing interest in using Green's function recovery technique to study the temperature effect on subsurface velocity variation in the lunar environment [18], to improve the near-surface ultrasonic array imaging resolution [19], etc. By comparing diffuse waves before and after external perturbations, time-lapse monitoring such as stress change [20], temperature variation [21], and progressive damage [22] can be achieved locally from only one fixed transmitter-receiver pair. Meanwhile, cross-correlation techniques with different stability and computational costs [16] have been developed under various names (e.g., cross-spectral moving-window technique (CSMWT), doublet method, stretching method, coda wave interferometry (CWI)) in different research communities. In 2002, Snieder et al. [23] showed that, together with the sensitivity kernel, the CWI technique can be expanded to have a full-field velocity perturbation. Similarly, Rossetto et al. [24] introduced an innovative technique called Locadiff for weak change inspection including emerging defects. Prior knowledge of the materials or exact wave propagation distance is not necessary with this method, which could be important for anisotropic structures [25]. Locadiff has been applied successfully for crack localization in pre-cracked concrete specimens under four point bending [12] with a spatial resolution of a few centimeters. The correlation/decorrelation coefficients are influenced by crack- and deformation-induced geometry changes. On-site detection of three dimensional multiple pre-existing cracks is also realizable for an aeronautical wind tunnel [26]. Combing the Locadiff and CWI techniques [23] together, both the microstructure and velocity variation fields can be obtained to better understand the mechanical behavior of natural rock samples [27].

Because of its great potential for structural health monitoring and nondestructive evaluation in complex structures, diffuse wave inspection has been investigated for fiber reinforced composites as well. Zhu et al. [28] have used coda waves to determine the internal stress in a polymer composite. Livings et al. [29] have explored the sensitivity of diffuse wave correlation coefficients, amplitude spectrum, and phase spectrum under different fatigue cycles for unidirectional carbon fiber reinforced polymer composites ([90/45/-45/90$_6$]$_S$). Waveform variation is not only caused by the fatigue cycle number, but also by the gain change, couplant type, excitation source type, frequency, etc. Although diffuse wave analysis can detect fatigue induced micro-cracks theoretically, precautions should be taken for all these experimental factors. Patra et al. [30] have applied a modified stretching technique to evaluate the progressive damage state of woven carbon fiber composites online under high-cycle-low-load fatigue loading. The precursor damage index (PDI), which is defined as the cumulative sum of the stretch parameters, can indicate the local formation of micro-scale defects. The sudden slope change of the PDI represents the stress state change from concentration to relaxation during defect generation. Recently, Pascal et al. [31] applied coda wave interferometry to monitor micro crack propagation in a polypropylene sulfide based carbon fiber composite with a layering sequence of [0°/90°, -45°/45°, 0°/90°, -45°/45°]$_s$ during a four point bending test. The relative velocity evolution can be derived from waveform correlations, which indicates the damage state.

To date, few attempts have been made to investigate defect localization in fiber reinforced composites with diffuse wave inspection. The present work aims to understand this procedure experimentally for a single defect localization in a cross-ply fiber reinforced structure. We will first review the Locadiff technique based on diffuse wave field. Then, the experiment results will be presented and analyzed.

2. Theoretical background

During diffuse wave propagation in a non-homogeneous material, the energy $I(S, R, t)$ of a diffuse wave in a defined frequency for a source I_0 can be described by the diffusion equation [32]:

$$\frac{\partial I(S, R, t)}{\partial t} - D\Delta I(S, R, t) + kI(S, R, t) = I_0, \tag{1}$$

where D is the wave diffusivity, k is the dissipation parameter, t is the time, $I(S, R, t)$ is the energy propagated from a source S to a receiver R. Equation (1) describes the spatio-temporal variation of the diffuse wave field. A more accurate description can be obtained from the radiative transfer equation [33].

Supposing the waveforms $\varphi_A(S, R, t)$ and $\varphi_B(S, R, t)$ can be obtained before and after a defect appearance during experiments, they are related with the energy through Equation (2) [24,34]:

$$\langle \varphi_A(S, R, t)\varphi_B(S, R, t) \rangle = I(S, R, t) - \frac{c\sigma}{2}\int_0^t I(S, x, u)I(x, R, t-u)du, \tag{2}$$

in which c is the wave velocity, σ is the scattering cross-section, and x is the defect location. It can be normalized into Equation (3):

$$\frac{\langle \varphi_A(S,R,t)\varphi_B(S,R,t) \rangle}{\sqrt{\langle \varphi_A^2(S,R,t) \rangle \langle \varphi_B^2(S,R,t) \rangle}} = CC(S, R, x, t) = 1 - DC(S, R, x, t) = 1 - \frac{c\sigma}{2}\frac{\int_0^t I(S,x,u)I(x,R,t-u)du}{I(S,R,t)}, \tag{3}$$

in which $CC(S, R, x, t)$ is the correlation coefficient and $DC(S, R, x, t)$ is the decorrelation coefficient. They all depend on the time and the positions of the source, the receiver, and the defect location. $DC(S, R, x, t) = 0$ when no defect presents ($\sigma = 0$) and $DC(S, R, x, t) = 1$ when the two waveforms are absolutely different (e.g., with large cracks). The decorrelation coefficient is related to the defect sensitivity kernel function $K(S, R, x, t)$ using Equations (4) and (5) [35]:

$$DC(S, R, x, t) = \frac{c\sigma}{2}K(S, R, x, t), \tag{4}$$

in which

$$K(S, R, x, t) = \frac{\int_0^t I(S, x, u)I(x, R, t-u)du}{I(S, R, t)} \tag{5}$$

Equation (5) does not have an analytical form in general and should be calculated numerically [33].

Once $DC(S, R, x, t)$ and $K(S, R, x, t)$ are obtained, the defect location can be predicted from different inversion algorithms, e.g., the linear least square inversion method [35] and the Monte Carlo Markov chain method [36]. Here a classical grid search method is used to find the most likely defect position using the cost function below:

$$e(x) = \sum_{S,R} \frac{DC(S, R, x, t)^2}{\varepsilon^2} - \frac{(\sum_{S,R} DC(S, R, x, t)K(S, R, x, t))^2}{\varepsilon^2 \sum_{S,R} K(S, R, x, t)^2}, \tag{6}$$

The probability density of the defect appearance at x is defined in Equation (7):

$$p(x) = \frac{1}{C}\exp(-\frac{e(x)}{2\varepsilon^2}), \tag{7}$$

where ε is a fluctuation parameter for the measured decorrelations and C is a normalization constant.

3. Experiments

A $[0°/90°]_{12}$ carbon fiber reinforced epoxy composite laminate made using a hot press process with unidirectional prepregs (Toray) is studied here; the experimental set-up is shown in Figure 1a. The plate is 510 mm × 510 mm × 3 mm in dimension. Since the thickness is small compared to the length and width dimensions, we will treat the plate as a two-dimensional model based on Equation (1). Due to the high attenuation of ultrasonic waves in the composite plate (4.3 dB/cm at 5 MHz [37]), the pitch-catch configuration is chosen and the gain is 17 dB. The high-pass filter is 1 MHz, while the low-pass filter is 10 MHz. The longitudinal transducer (A551S 5 MHz, Olympus, Inc., Tokyo, Japan) is excited using a pulsed source from a signal generator (Shantou Goworld Display Co., Ltd., Shantou, China). The waveform is recorded using a same type longitudinal transducer through an oscilloscope (TBS1202B, Tektronix, Inc., Beaverton, OR, USA) connected to a personal computer. The ultrasonic sensors are fixed to the plate by 3D printed fixtures with springs inside to apply constant forces on the sensors. A high viscosity couplant can provide good wave transmission and a traditional honey is applied. The defect is simulated using a circular piezoelectric medium attached to the composite plate surface [38]. Signals are amplified and averaged over 10 acquisitions for each source-receiver pair to minimize noises such as the rapid changes of capacitances between conductors due to flexing, twisting, or transient impacts on cables. Furthermore, since an important signal variation has been observed with a minor sensor movement, the defect is attached and detached repeatedly while keeping each source (S)-receiver (Rn, n = 1–4) pair fixed during signal recording (Figure 1b).

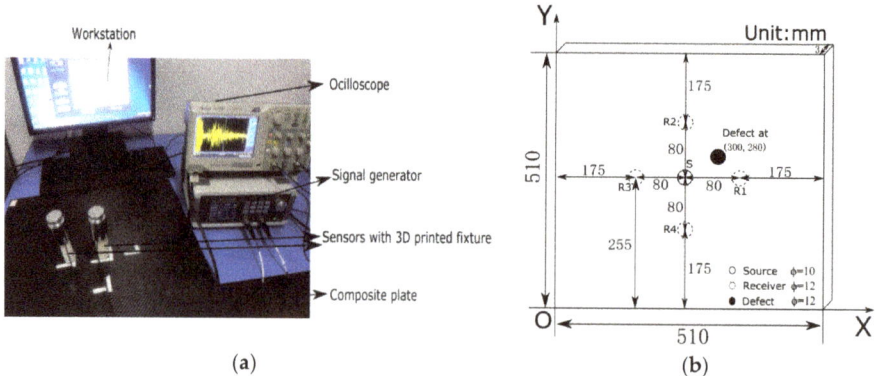

Figure 1. (a) Experimental set-up and (b) measurement scheme.

A typical waveform from S-R_1 is presented in Figure 2. No first arrival time variation can be observed using an amplitude threshold picker [39]. The wave diffusivity and the dissipation parameter can be decided from the waveform envelope through the Hilbert transform based on Equation (1) that $D = 100$ m^2/s and $k = 3 \times 10^5$ /s. The dissipation parameter k is an indication of viscoelasticity, which describes the exponential decay at late times. It is much larger than that of concrete with 5500/s [26]. These values fit well with waveforms in 0° and 90° directions. In contrast, the waveforms from off-axis directions (30° and 60°) have poor signal-to-noise ratios, as shown in Figure 3. It can be difficult for the diffusion property to be decided in those directions, which makes measurement with an arbitrary sensor location unreliable. Only the measurements along the X and Y axes are chosen here, and a constant diffusion property is assumed during the defect sensitivity kernel calculation under the measurement scheme in Figure 1b. The first arrival time is about 10 µs and gives a group velocity of 8000 m/s along the X and Y axes. In fact, the directivity of the group velocity [40] does not interfere with the defect localization procedure according to Equation (6). The transport mean free path $L^* = 2D/c \approx 0.025$ m is less than the transmitter-receiver pair distance 0.08 m, which ensures the ultrasonic wave is multiple scattered. The wave diffusivity is a characteristic of the microstructure which relates to the arrival time of the maximum energy density and decreases with frequency. A large variation of D has been found for concrete, from 10 m^2/s [32] to 70 m^2/s [26], because of its internal structure variation and frequency sensitivity to the diffuse wave. This can be expected as well for fiber reinforced composites. Quiviger et al. found that D varies from 17 to 10 m^2/s, while the crack size increases from 1cm to 5.5 cm [32]. However, the defect generated here is weak enough that D remains unchanged. This mimics an early stage surface crack initiation.

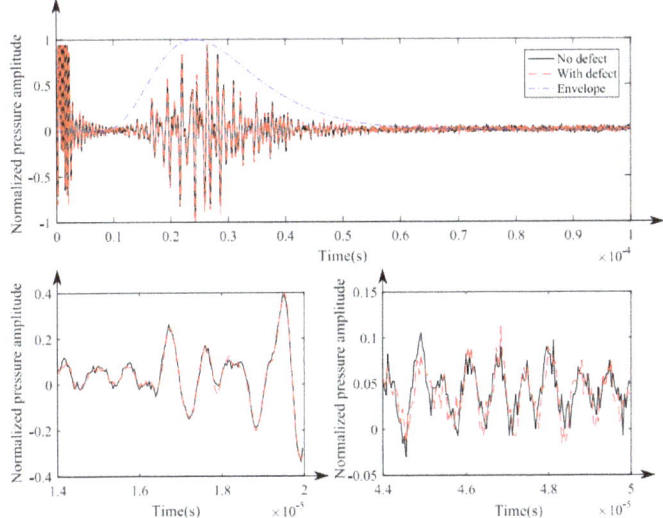

Figure 2. Representative waveforms from the S-R1 pair.

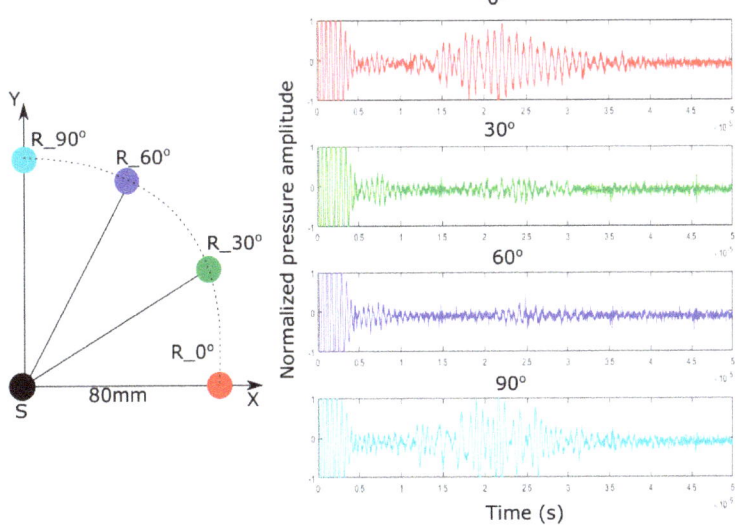

Figure 3. Waveforms from different directions at 0°, 30°, 60°, and 90°.

4. Results and Discussion

4.1. Decorrelation Coefficient and Defect Sensitive Window

According to Equation (3), the correlation and decorrelation coefficients can be determined from the waveforms with and without a defect. The stretching method demonstrates great stability to external noise and is applied here [16]. The decorrelation coefficients can be calculated according to Equation (8):

$$DC(\varepsilon) = 1 - CC(\varepsilon) = 1 - \frac{\int_{t_1}^{t_2} h_k[t(1-\varepsilon)]h_0[t]dt}{\sqrt{\int_{t_1}^{t_2} h_k^2[t(1-\varepsilon)]dt \int_{t_1}^{t_2} h_0^2[t]dt}}, \quad (8)$$

in which the window length $T = t_2 - t_1$ is set to 10 µs and the beginning of the time window t_1 is shifted sequentially with a time step of 5 µs each time. The stretching factor ε is chosen to vary from −0.015 to 0.015, from which the decorrelation coefficient is decided when reaching the minimum (Figure 4). The total results for all decorrelation coefficients varied with time are shown in Figure 5. They all increase with time as a growing interaction between the diffuse wave field and the defect takes place. The maximum coefficients are found when $t_1 = 45$ µs, which is chosen for the defect localization study. S-R1 and S-R2 are more sensitive to the defect compared to that of S-R3 and S-R4 due to their different positions relative to the defect. After $t_1 = 45$ µs, the decorrelation coefficients decrease because of the reflection part interference and signal attenuation.

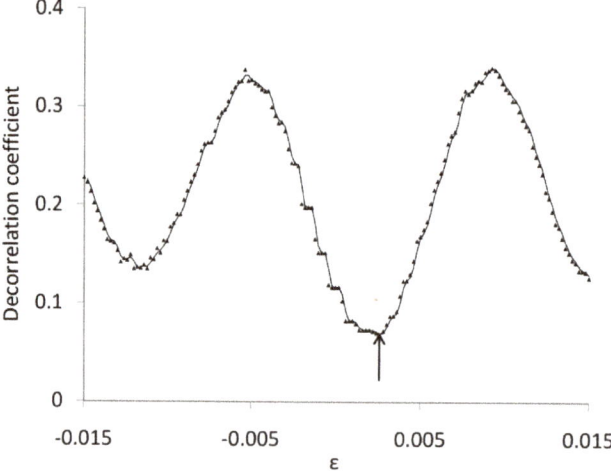

Figure 4. Decorrelation coefficient calculation example for the waveforms presented in Figure 2 when $t_1 = 45$ µs (decorrelation coefficient = 0.0679 when $\varepsilon = 0.0026$).

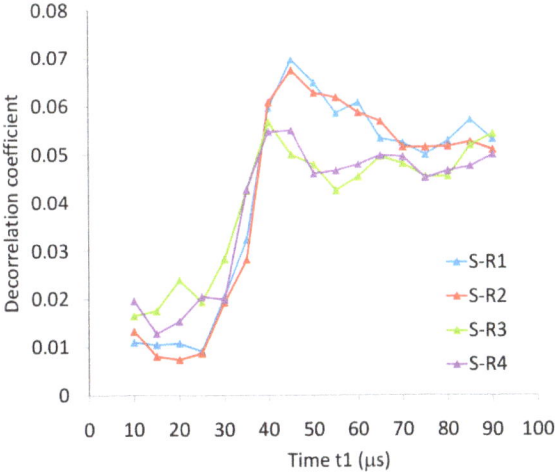

Figure 5. Decorrelation coefficients varied according to time.

4.2. Kernel Function

The defect sensitivity kernel represents the probability of a wave sent from location S to pass at location x and then to arrive at location R after period of time t [11]. Considering the source images due to the four straight boundaries of the plate, the detection diameter $Z_{detection}$ is 260 mm when $t = 45$ μs, as shown in Figure 6 based on Equation (5). The diffuse wave field is mostly sensitive to the defect in this range. In fact, the diffusion model simplifies the energy distribution and neglects the diffuse field-defect interaction after $t = 65$ μs, as previously shown in Figure 2. All the pressure amplitudes are equal to zero in that period. The whole space is then discretized into 300 × 300 elements for the possible location of the defect, and a further mesh refinement gives no more localization resolution improvement.

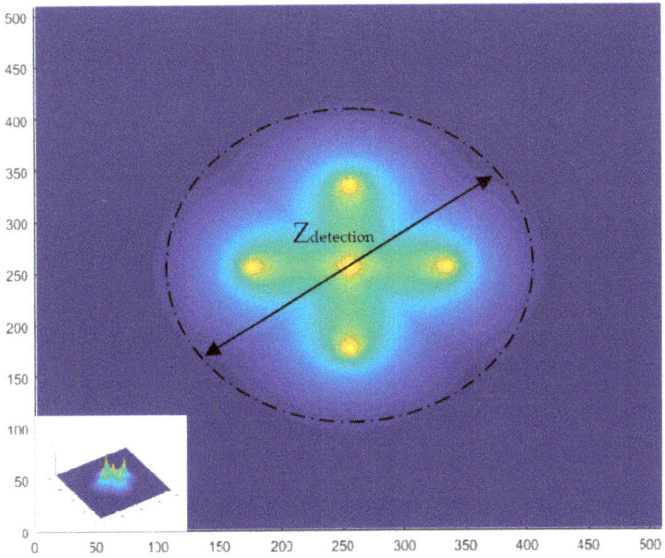

Figure 6. Sensitivity kernels and detection zone.

4.3. Defect Localization and Error Discussion

Once the decorrelation coefficients and kernel function have been calculated in the defined time domain, the localization is found to be in [270 mm, 265 mm] based on Equations (4)–(7), as shown in Figure 7. A further comparison is made among different time windows in Figure 8. The localization error varies from 13.5% to 6.7%. It decreases initially due to the increasing interaction between the diffuse wave and the defect. The direct reflection wave is estimated to arrive at the defect position at $t_{reflection} = \frac{255+(255-80)}{8000} = 53$ μs. Subsequently, the localization error increases because of the lowering sensitivity of the waveform to the defect shown in Figure 5, which may be caused by the reflection and attenuation. Although they suffer a lot from attenuation, which limits the inspection domain, a meaningful signal can still be extracted.

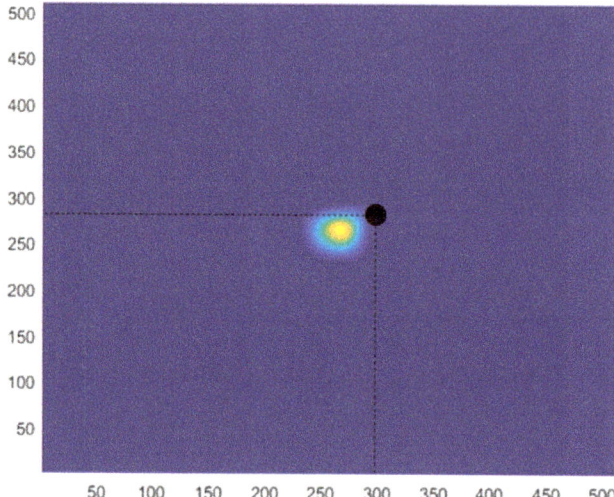

Figure 7. Localization image when t = 45 us.

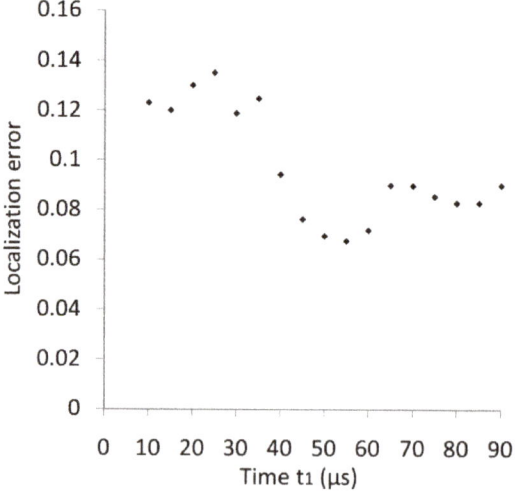

Figure 8. Localization error varied with time window.

In addition to the severe signal-to-noise ratio degradation due to viscoelasticity, the moderate defect localization deviation mainly comes from two aspects. Theoretically, wave diffusivity and dissipation parameters are assumed constant in the kernel calculation model, which varies according to the directional angles. This approximation contributes to the error in the kernel calculation and grid search procedure in Equation (6). Experimentally, the signal obtained from each pair is the averaged information from the sensor's field of view. Localization sensitivity may be lost to some extent so that error is introduced as well. The sensor diameter and distribution could be optimized in the future between the direction sensitivity and inspection efficiency in order to obtain high defect localization precision.

5. Conclusions

Fiber reinforced polymer composites can possess complex structures from macroscopic to microscopic scales, which makes their quality assurance difficult. Diffuse wave inspection relies on the multiple scattering process containing rich structure information. It gives a new insight into high frequency ultrasonic inspection for such viscoelastic, inhomogeneous, and anisotropic structures. The diffuse wave field is attenuated fast, while the decorrelation coefficient decreases first with increasing sensitivity to the defect and increases later because of signal-to-noise ratio degradation. Wave diffusivity is assumed constant under the measurement scheme only along the X and Y axes. The most defect sensitive time window is decided in the range of 45–55 µs. Limited by the sensor diameter and waveform attenuation, the defect is found to locate at [270 mm, 265 mm] compared to [300 mm, 280 mm] in the intact state in the given window. This method is promising for early crack detection in various advanced composites in three dimensions. Since the diffuse wave is the statistical summation of random walk ray-paths [23], more experimental studies are expected for different fiber structures. Further investigation of how to distinguish signals from noises in diffuse waves is expected in viscoelastic media, which could be critical for defect localization.

Author Contributions: Data curation, Y.D.; formal analysis, D.T.; investigation, H.Z.; writing—original draft, Q.Z.; writing—review and editing, Y.P.

Funding: The work in this paper is supported by the Shanghai Sailing Program [No.18YF1408400] and the National Natural Science Foundation of China (Grant Nos.116742124, 11874255, 61673252).

Conflicts of Interest: The authors declare no conflicts of interest.

References

1. Llewellyn-Jones, T.M.; Drinkwater, B.W.; Trask, R.S. 3D printed composites with ultrasonically arranged complex microstructure. *Smart Mater. Struct.* **2016**, *25*, 02LT01. [CrossRef]
2. Erb, R.M.; Libanori, R.; Rothfuchs, N.; Studart, A.R. Composites reinforced in three dimensions by using low magnetic fields. *Science* **2012**, *335*, 199–204. [CrossRef] [PubMed]
3. Raney, J.R.; Compton, B.G.; Mueller, J.; Ober, T.J.; Shea, K.; Lewis, J.A. Rotational 3D printing of damage-tolerant composites with programmable mechanics. *PNAS* **2018**, *115*, 1198–1203. [CrossRef] [PubMed]
4. Lascoup, B.; Perez, L.; Autrique, L. Defect localization based on modulated photothermal local approach. *Compos. Part B Eng.* **2014**, *65*, 109–116. [CrossRef]
5. Yang, B.; Xuan, F.Z.; Chen, S.; Zhou, S.; Gao, Y.; Xiao, B. Damage localization and identification in WGF/epoxy composite laminates by using Lamb waves: Experiment and simulation. *Compos. Struct.* **2017**, *165*, 138–147. [CrossRef]
6. Pearson, M.R.; Eaton, M.; Featherston, C.; Pullin, R.; Holford, K. Improved acoustic emission source location during fatigue and impact events in metallic and composite structures. *Struct. Heal. Monit.* **2017**, *16*, 382–399. [CrossRef]
7. Kalteremidou, K.-A.; Murray, B.; Tsangouri, E.; Aggelis, D.; Van Hemelrijck, D.; Pyl, L. Multiaxial damage characterization of carbon/epoxy angle-ply laminates under static tension by combining in situ microscopy with acoustic emission. *Appl. Sci.* **2018**, *8*, 2021. [CrossRef]

8. Niccolini, G.; Borla, O.; Accornero, F.; Lacidogna, G.; Carpinteri, A. Scaling in damage by electrical resistance measurements: An application to the terracotta statues of the Sacred Mountain of Varallo Renaissance Complex (Italy). *Rend. Lincei* **2015**, *26*, 203–209. [CrossRef]
9. Chen, B.; Liu, J. Damage in carbon fiber-reinforced concrete, monitored by both electrical resistance measurement and acoustic emission analysis. *Constr. Build. Mater.* **2008**, *22*, 2196–2201. [CrossRef]
10. Snieder, R. Time-reversal invariance and the relation between wave chaos and classical chaos. In *Imaging of Complex Media with Acoustic and Seismic Waves*; Flink, M., Kuperman, W.A., Montagner, J.-P., Tourin, A., Eds.; Springer: Berlin, Germany, 2002; pp. 1–16.
11. Xie, F.; Larose, E.; Moreau, L.; Zhang, Y.; Planes, T. Characterizing extended changes in multiple scattering media using coda wave decorrelation: Numerical simulations scattering media. *Waves Random Complex Media* **2018**, *5030*, 1–14. [CrossRef]
12. Larose, E.; Obermann, A.; Digulescu, A.; Planès, T.; Chaix, J.-F.; Mazerolle, F.; Moreau, G. Locating and characterizing a crack in concrete with diffuse ultrasound: A four-point bending test. *J. Acoust. Soc. Am.* **2015**, *138*, 232–241. [CrossRef] [PubMed]
13. Biswal, P.; Reddy, B.N.; Srinivasa, M.B.P. Manufacturing aspects of fabrication of composite reference standard for NDT ultrasonic inspection. In Proceedings of the AeroNDT, Bangalore, India, 3–5 November 2016.
14. Hennings, B.; Lammering, R. Material modeling for the simulation of quasi-continuous mode conversion during Lamb wave propagation in CFRP-layers. *Compos. Struct.* **2016**, *151*, 142–148. [CrossRef]
15. Aki, K.; Chouet, B. Origin of coda waves:Source, attenuation, and scattering. *J. Geophys. Res.* **1975**, *80*, 3322–3342. [CrossRef]
16. Hadziioannou, C.; Larose, E.; Coutant, O.; Roux, P.; Campillo, M. Stability of monitoring weak changes in multiply scattering media with ambient noise correlation: Laboratory experiments. *J. Acoust. Soc. Am.* **2009**, *125*, 3688–3695. [CrossRef] [PubMed]
17. Lobkis, O.I.; Weaver, R.L. On the emergence of the Green's function in the correlations of a diffuse field. *Ultrasonics* **2001**, *110*, 3011–3017. [CrossRef]
18. Sens-Schonfelder, C.; Larose, E. Lunar noise correlation, imaging and monitoring. *Earthq. Sci.* **2010**, *23*, 519–530. [CrossRef]
19. Potter, J.N.; Wilcox, P.D.; Croxford, A.J. Diffuse field full matrix capture for near surface ultrasonic imaging. *Ultrasonics* **2018**, *82*, 44–48. [CrossRef] [PubMed]
20. Stahler, S.C.; Sens-Schonfelder, C.; Niederleithinger, E. Monitoring stress changes in a concrete bridge with coda wave interferometry. *J. Acoust. Soc. Am.* **2011**, *129*, 1945–1952. [CrossRef]
21. Zhang, Y.; Abraham, O.; Tournat, V.; Le Duff, A.; Lascoup, B.; Loukili, A.; Grondin, F.; Durand, O. Validation of a thermal bias control technique for Coda Wave Interferometry (CWI). *Ultrasonics* **2013**, *53*, 658–664. [CrossRef]
22. Shokouhi, P. Stress- and damage-induced changes in coda wave velocities in concrete. In Proceedings of the AIP Conference Proceedings, Denver, CO, USA, 15–20 July 2012.
23. Snieder, R. The Theory of Coda Wave Interferometry. *Pure Appl. Geophys.* **2006**, *163*, 455–473. [CrossRef]
24. Rossetto, V.; Margerin, L.; Planes, T.; Larose, E. Locating a weak change using diffuse waves: Theoretical approach and inversion procedure. *J. Appl. Phys.* **2011**, *109*, 034903. [CrossRef]
25. Kundu, T.; Yang, X.; Nakatani, H.; Takeda, N. A two-step hybrid technique for accurately localizing acoustic source in anisotropic structures without knowing their material properties. *Ultrasonics* **2015**, *56*, 271–278. [CrossRef] [PubMed]
26. Zhang, Y.; Larose, E.; Moreau, L.; d'Ozouville, G. Three-dimensional in-situ imaging of cracks in concrete using diffuse ultrasound. *Struct. Heal. Monit.* **2018**, *17*, 279–284. [CrossRef]
27. Xie, F.; Ren, Y.; Zhou, Y.; Larose, E.; Baillet, L. Monitoring local changes in granite rock under biaxial test: A spatiotemporal imaging application with diffuse waves. *J. Geophys. Res. Solid Earth* **2018**, *123*, 2214–2227. [CrossRef]
28. Zhu, Q.; Binetruy, C.; Burtin, C. Internal stress determination in a polymer composite by Coda wave interferometry. In Proceedings of the IOP Conference Series: Materials Science and Engineering, Hangzhou, China, 20–23 May 2016.
29. Livings, R.; Dayal, V.; Barnard, D. Feasibility of detecting fatigue damage in composites with Coda waves. In Proceedings of the AIP Conference Proceedings, Boise, ID, USA, 20–25 July 2014.

30. Patra, S.; Banerjee, S. Material state awareness for composites part I: Precursor damage analysis using ultrasonic guided coda wave interferometry (CWI). *Materials* **2017**, *10*, 1436. [CrossRef] [PubMed]
31. Pomarède, P.; Chehami, L.; Declercq, N.F.; Meraghni, F.; Dong, J.; Locquet, A.; Citrin, D.S. Application of ultrasonic Coda wave interferometry for micro-cracks monitoring in woven fabric composites. *J. Nondestruct. Eval.* **2019**, *38*, 26. [CrossRef]
32. Quiviger, A.; Payan, C.; Chaix, J.; Garnier, V.; Salin, J. Characterizing extended changes in multiple scattering media using coda wave decorrelation. *NDT E Int.* **2012**, *45*, 128–132. [CrossRef]
33. Kanu, C.; Snieder, R. Numerical computation of the sensitivity kernel for monitoring weak changes with multiply scattered acoustic waves. *Geophys. J. Int.* **2015**, *203*, 1923–1936. [CrossRef]
34. Planeès, T.; Larose, E.; Rossetto, V.; Margerin, L. LOCADIFF: Locating a weak change with diffuse ultrasound. In Proceedings of the AIP Conference Proceedings, Denver, CO, USA, 15–20 July 2012.
35. Planès, T.; Larose, E.; Rossetto, V.; Margerin, L. Imaging multiple local changes in heterogeneous media with diffuse waves. *J. Acoust. Soc. Am.* **2015**, *137*, 660–667. [CrossRef] [PubMed]
36. Xie, F.; Moreau, L.; Zhang, Y.; Larose, E. A Bayesian approach for high resolution imaging of small changes in multiple scattering media. *Ultrasonics* **2016**, *64*, 106–114. [CrossRef] [PubMed]
37. Williams, J.H.; Nayeb-Hashemi, H.; Lee, S.S. Ultrasonic attenuation and velocity in AS/3501-6 graphite fiber composite. *J. Nondestruct. Eval.* **1980**, *1*, 137–148. [CrossRef]
38. Chehami, L.; Moulin, E.; De Rosny, J.; Prada, C.; Assaad, J.; Benmeddour, F. Experimental study of passive defect detection and localization in thin plates from noise correlation. *Phys. Procedia* **2015**, *70*, 322–325. [CrossRef]
39. Carpinteri, A.; Xu, J.; Lacidogna, G.; Manuello, A. Reliable onset time determination and source location of acoustic emissions in concrete structures. *Cem. Concr. Compos.* **2012**, *34*, 529–537. [CrossRef]
40. Mei, H.; Haider, M.F.; Joseph, R.; Migot, A.; Giurgiutiu, V. Recent advances in piezoelectric wafer active sensors for structural health monitoring applications. *Sensors* **2019**, *19*, 383. [CrossRef] [PubMed]

© 2019 by the authors. Licensee MDPI, Basel, Switzerland. This article is an open access article distributed under the terms and conditions of the Creative Commons Attribution (CC BY) license (http://creativecommons.org/licenses/by/4.0/).

Article

Assessment of Residual Elastic Properties of a Damaged Composite Plate with Combined Damage Index and Finite Element Methods

Carlo Boursier Niutta *, Andrea Tridello, Raffaele Ciardiello, Giovanni Belingardi and Davide Salvatore Paolino

Department of Mechanical and Aerospace Engineering, Politecnico di Torino, 10129 Turin, Italy; andrea.tridello@polito.it (A.T.); raffaele.ciardiello@polito.it (R.C.); giovanni.belingardi@polito.it (G.B.); davide.paolino@polito.it (D.S.P.)
* Correspondence: carlo.boursier@polito.it

Received: 29 May 2019; Accepted: 21 June 2019; Published: 25 June 2019

Abstract: In structural component applications the use of composite materials is increasing thanks to their optimal mechanical characteristics. However, the complexity of the damage evolution in composite materials significantly limits their widespread diffusion. Non-destructive tests are thus becoming ever more important. The detecting Damage Index (DI_d) technique has been recently brought in the realm of the non-destructive characterization tests for components made of composite material. In contrast to other techniques, this methodology allows to quantitatively assess local residual properties. In this paper, the DI_d technique is adopted in combination with the finite element method. The mechanical response of two composite plates (an 8-layer twill fabric carbon/epoxy) subjected to four-point bending test is firstly used to tune a finite element model of the laminate. Then, an undamaged laminate of the same composite material is progressively damaged through repeated four-point bending tests. Local residual elastic properties are mapped on the plate through the DI_d technique. A continuous polynomial curve has been considered to account for the variation of the elastic modulus in the finite element model. The resulting force-displacement curve of the numerical analysis is compared to experimental data of damaged plate, resulting in very good agreement. The combination of the experimental activity and the numerical finite element analysis points out the accuracy of the DI_d methodology in assessing local residual elastic properties of composite materials.

Keywords: non-destructive tests; damage assessment; residual properties; Finite Element Method; Damage Index

1. Introduction

Composite materials are increasingly adopted in structural–mechanical applications thanks to their optimal characteristics in terms of light weight, mechanical strength and stiffness, corrosion resistance, energy absorption capacity, and noise attenuation. However, many factors limit their widespread diffusion, such as the high cost of raw materials or requirements for high production volumes. Among the others, the complexity of the damage evolution plays a key role in their limitation, especially for structural component applications. Several interacting failure modes are typical of composite materials and a progressive and rapid decrement of the mechanical properties can be observed [1].

In this regard, methodologies for assessing damage level and predicting the residual structural strength of composite materials are becoming increasingly important. Many techniques have been developed for non-destructively assessing the structural health state of composite components [2]. Non-destructive tests (NDTs) are commonly adopted for quality assessment of a manufacturing process or for damage evaluation in structures during service. Most of the techniques intend to provide information on internal damages, in terms of size, shape, location, and orientation. Microscopy (based on atomic force, optic system, and scanning and transmission electron) [3], X-ray micro-CT (micro-computed tomography) [3,4], and infrared thermography [3,5] are among the most used for composite applications. However, the qualitative information provided by these techniques can be exploited by designers with difficulty. Further, methods based on ultrasounds and vibrational analysis are usually adopted for globally analyzing the dynamic response of a structure [6]. In particular, the use of the Impulse Excitation Technique (IET), which is regulated by ASTM Standards for metallic materials [7,8], has been recently extended to composite specimens [9]. However, local damages and defects are still revealed with difficulty in real-world structures since the presence of anomalies is mitigated by the global response of the system. Therefore, as pointed out in [10], current methodologies do not allow a direct evaluation of the local residual properties.

In this paper, the detecting Damage Index (DI_d) technique was adopted for assessing the local residual elastic properties of a composite laminate. This methodology was brought in the realm of the non-destructive characterization for components made of composite material by Belingardi et al. in [11–15]. The DI_d technique consists of two sets of experimental tests: a preliminary characterization activity which is performed to identify the correlation between residual elastic properties of the material and the DI_d parameter and the proper non-destructive test to estimate the local residual elastic properties of the investigated component from the DI_d. In [15], the methodology was demonstrated to properly predict residual elastic properties of a series of laminate plates damaged through impacts at different energy levels. Tensile tests on specimens cut from laminate plates allowed validation of results predicted with the DI_d. Here, the finite element method was adopted to validate elastic properties assessed with the DI_d. The combination of non-destructive tests and finite element method is typical in damage assessment, particularly in vibration-based techniques [16]. Various approaches are available to numerically model multi-layered composite materials and their multiple failure modes [17,18]. In particular, the use of cohesive elements disposed between layers of shells currently represents one of the most adopted solutions [19–21]. Thanks to their specific formulation, cohesive elements allows to simulate damages due to delamination without significantly affecting computational cost.

In this paper, the DI_d technique is thus used in combination with the finite element method. An 8 layers twill fabric carbon/epoxy composite laminate is firstly damaged through repeated four-point bending tests. Then, the DI_d technique is adopted to map the residual elastic properties on the damaged plate. The variation of the Young's modulus is accounted in the finite element model of the laminate subjected to bending test through a continuous polynomial curve. The resulting experimental and numerical force-displacement curves are then compared to validate the proposed methodology. The combination of the experimental activity and the numerical finite element analysis points out the accuracy and the effectiveness of the DI_d methodology in predicting the local residual elastic properties of damaged composite components.

2. Materials and Methods

In this section, investigated materials are firstly detailed. Then experimental and numerical methods are presented.

2.1. Materials

Experimental tests are performed on a structural composite laminate specifically developed for automotive application. The composite material is the same adopted by Tridello et al. in their investigations [15] and previously in [9]. It consists of a matrix made of epoxy resin reinforced by eight layers of twill woven carbon fabric. The first layer is a 380 gsm fabric with 0.45 mm thickness and an 800 gsm fabric with 0.88 mm thickness is used in the remaining seven layers. The stacking sequence is $[0/90]_8$. In the numerical analysis, the laminate is modeled as symmetric, with 8 layers of thickness 0.88 mm, oriented according to the stacking sequence. A total thickness of 7.04 mm is thus obtained. Even though the first layer is different from the other seven, this difference can be neglected when considering the global mechanical behavior of the laminate. The measured elastic properties (Young's modulus, shear modulus, and Poisson's ratio) are reported in Table 1, as assessed in [9]. Given the symmetry of the woven fabric, the mechanical properties and the Poisson's ratios in the in-plane direction 1 and 2 are the same (i.e. $E_1 = E_2$, $\nu_{12} = \nu_{21}$).

Table 1. Material properties.

Property	Value
Density	1.45 10^3 g/cm^3
Young's modulus in longitudinal and transverse direction ($E_1 = E_2$)	54 GPa
Shear modulus (G_{12})	3.5 GPa
Poisson's ratio ($\nu_{12} = \nu_{21}$)	0.08

2.2. Experimental Tests

Four-point bending tests are performed on two different composite plates until complete failure occurs. The elastic field of the composite plates is thus identified. The tests are performed on a servohydraulic testing machine (Instron 8801). In agreement with the recommendations of ASTM standard D6272 for four-point bending test of reinforced plastics [22], the load span is one third of the support span. As shown in Figure 1, the loading noses are 50 mm distant. The support noses are consequently placed at a distance of 150 mm. Figure 1a shows the experimental setup. The resulting experimental curves of force with respect to displacement are exploited to tune a finite element model of the laminate subjected to four-point bending test, which is shown in Figure 1b.

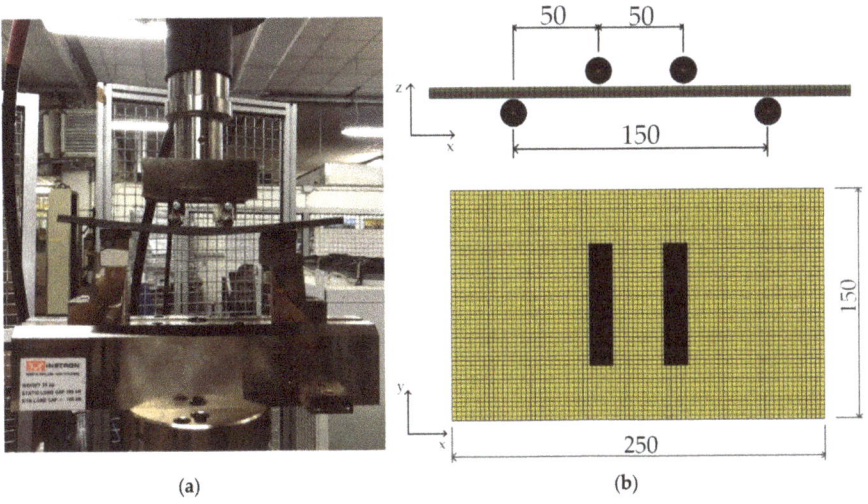

Figure 1. Four-point bending test: (a) experimental setup; (b) numerical model.

The same experimental setup is adopted to progressively damage an undamaged composite plate. The plate is loaded until incipient failure occurs and then unloaded. The repeated four-point bending tests progressively reduce plate stiffness. A significant damage level is thus induced.

In order to assess the residual Young's modulus of the damaged plate, the detecting Damage Index (DI_d) technique is adopted. The DI_d methodology consists of two sets of experimental tests. Firstly, a preliminary characterization activity is performed, which allows to identify the correlation between residual elastic properties and the DI_d parameter. Then, the investigated component undergoes the non-destructive test in order to estimate the local residual elastic properties from the DI_d.

In particular, the methodology is based on two sets of impact tests: the first intends to damage the material at increasing impact energy levels and allows to evaluate the so-called threshold energy ε_{th}. The threshold energy is defined as the impact energy at which the reduction of local elastic properties is less than 5%. An impact at the threshold energy can be considered as non-destructive for the material. The second set of impacts is performed at the threshold energy and allows to evaluate the residual elastic properties.

In this work, the first series of impacts were performed on a set of plates made of the same composite fabric. The impact tests were carried out using a free-fall drop dart testing machine (CEAST 9350 FRACTOVIS PLUS). Clamping boundary conditions were realized through a mechanical clamping system which applies an almost uniform pressure on the clamped area. A circular unclamped region of diameter 76 mm was considered for the tests, in agreement with the recommendations of ASTM standard D5628 [23]. The impact energy was defined by varying the impactor mass for a given impact velocity. The impact velocity, which was controlled by the drop height of the dart, was measured in each test with an optoelectronic device. A piezoelectric load cell, mounted in proximity of the tip of the impact dart, acquired the force signal at a sample rate of 1 MHz.

The second series of impacts, which are carried out at the threshold energy, is performed on the plate previously damaged with the repeated four-point bending tests. In particular, damaged plate is impacted along the middle line. Moving along the longitudinal direction of the plate (x direction of Figure 1), the impacts are located in correspondence of the external noses, in the regions where bending moment is linearly increasing, in correspondence of the internal noses and finally in the middle of the load span. Seven impacts are applied in total to the plate, as shown in Figure 2. The corresponding residual Young's modulus is calculated from the DI_d parameter, which is defined as

$$DI_d = \frac{\varepsilon_a}{\varepsilon_{th}} \cdot \frac{s_{MAX}}{s_{QS}}$$

where ε_a is the absorbed energy, ε_{th} the threshold energy, s_{MAX} the maximum displacement and s_{QS} the displacement obtained in quasi static perforation test. The s_{MAX} and ε_a values are computed through numerical integrations of the load signal acquired during the impact test.

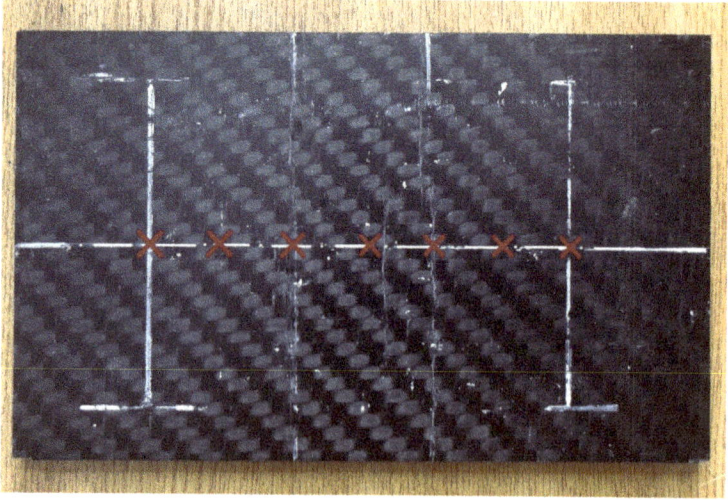

Figure 2. Location of impacts for DI_d analysis.

2.3. Numerical Model

A structural–mechanical model of the four-point bending test was simulated using the commercial software LS-DYNA. The 8 layers of the woven fabric composite were modeled with Belytschko-Tsai 4-node shell elements, with one integration point through the thickness. For each layer of the composite, a layer of shell elements was considered. Solid elements of cohesive material were placed between the shell layers, in order to represent the resin among layers. Solid elements of cohesive consider a specific formulation of LS-DYNA, which allows the transfer of moments to the shells. In particular, nodes of solid elements coincide with those of the shells. The material card *MAT_LAMINATED_COMPOSITE_FABRIC, which is specific for fabric composite, was adopted for the composite layers. This material model assumes a slightly modified version of Hashin criteria for failure [18]. However, in this paper, the focus is on the elastic field, and parameters related to the post-failure behavior of the material will not be addressed. The cohesive material model *MAT_COHESIVE_TH, with properties of the resin, was assigned to the cohesive elements. Finally, the span noses were modeled as rigid cylinder walls. The upper cylinders, shown in Figure 1b, move downwards with a prescribed motion law. The upper and lower cylinders were placed at the initial distance of 1 mm from the laminate in order to avoid penetrations and numerical instabilities. Contact between laminate and cylinders was based on a penalty formulation with a penalty factor which adaptively guarantees the numerical stability [24]. This formulation accounts for thickness offsets of shells.

The accuracy of the numerical model was firstly verified with respect to the four point-bending tests of the two composite plates. Then, the same numerical model was used in comparison with the experimental results of the progressively damaged plate. In this case, the structural–mechanical model had to account for the variation of Young's modulus on the plate, as pointed out through the DI_d technique.

The residual elastic properties of the damaged plate are mapped through the DI_d technique. The Young's modulus varies along the x-direction according to the location of measurement, with the damage mostly located in correspondence of internal rolls. It seems reasonable to assume that the Young's modulus varies continuously along the longitudinal direction. Here, a polynomial curve is considered for each region included between two consecutive measure locations.

For each polynomial curve, the two Young's moduli at the extremities, E_k and E_{k+1}, with k the considered location, are known.

In addition, the two derivatives at the extremities can be estimated. In order to evaluate the derivates, the difference quotient for each region has been calculated as

$$\frac{\Delta E_k}{\Delta x_k} = \frac{E_k - E_{k-1}}{x_k - x_{k-1}} \quad (1)$$

in accordance with its definition.

The difference quotient is a measure of the average rate of change of the function, here the Young's modulus, over the interval Δx_k. By considering two consecutive difference quotients, $\frac{\Delta E_k}{\Delta x_k}$ and $\frac{\Delta E_{k+1}}{\Delta x_{k+1}}$, the derivative at the k-th location has been estimated as follows:

1. when the product $\frac{\Delta E_k}{\Delta x_k} \cdot \frac{\Delta E_{k+1}}{\Delta x_{k+1}}$ returns a negative value, the derivate at the k-th location is assumed equal to zero. This can be justified by considering that a change in the sign of the difference quotient implies a change in the derivative of the function, as well;
2. when the sign of the product $\frac{\Delta E_k}{\Delta x_k} \cdot \frac{\Delta E_{k+1}}{\Delta x_{k+1}}$ is positive, the derivate at the k-th location is assumed equal to average value $\frac{\frac{\Delta E_k}{\Delta x_k} + \frac{\Delta E_{k+1}}{\Delta x_{k+1}}}{2}$;
3. the derivates at the extremities, $k = 1$ and $k = 7$, are assumed equal to zero.

As the two Young's moduli and the derivatives are known at the extremities, a third-order polynomial curve can be constructed for each interval included between two consecutive impact locations. This approach guarantees the continuity of the Young's modulus with respect to the x-coordinate.

In the structural–mechanical model, this variation was accounted by longitudinally dividing the layers of shells into parts. Each part consisted of one row of shell elements and is identified by its x-coordinate. In particular, each part could be longitudinally localized in correspondence of the middle of the element size. In this work, a mesh of 3 mm was adopted. According to the location of the part in the x direction, the corresponding third-order polynomial curve could be identified. The Young's modulus was consequently calculated. A material card was defined for each part with the corresponding elastic properties.

3. Results

In this Section, the experimental and numerical results are presented. In Section 3.1, experimental and numerical data related to the four-point bending tests are compared. In Section 3.2, results of the progressive damaging of the composite plate through the repeated four-point bending tests are shown. Residual elastic properties are thus estimated through the DI_d methodology. The Young's modulus is mapped with respect to the longitudinal coordinate and the third-order polynomial curves were constructed as previously described. Finally, in Section 3.3, the proposed approach is validated by comparing the experimental results of the progressively damaged plate with the numerical model which accounts for the residual properties.

3.1. Four-Point Bending Test: Experimental Results and Numerical Model Tuning

Results of the four-point bending tests are here presented. Two undamaged composite plates have been tested until almost complete failure occurred. In particular, the upper layers, which are subjected to compressive loads, showed the most significant failures. Intralaminar cracks propagate in correspondence of the inner span noses. The numerical model is analyzed only in the elastic field and it results in good agreement with the experimental data, as shown in Figure 3. Displacement of the numerical analysis has been evaluated on the rigid cylinders. However, as described in Section 2.2, the rigid wall cylinders are initially distanced from the composite laminate. Consequently, the initial path of the force-displacement curve is characterized by the growing contact between laminate and cylinders, which is not significant for the purpose. The numerical force-displacement curve has been thus shifted leftwards and results of the numerical analysis are reported from 1 mm of displacement.

Figure 4 shows a magnification of the cracks, obtained through an optical microscope. The complete failure can be appreciated for the upper four layers.

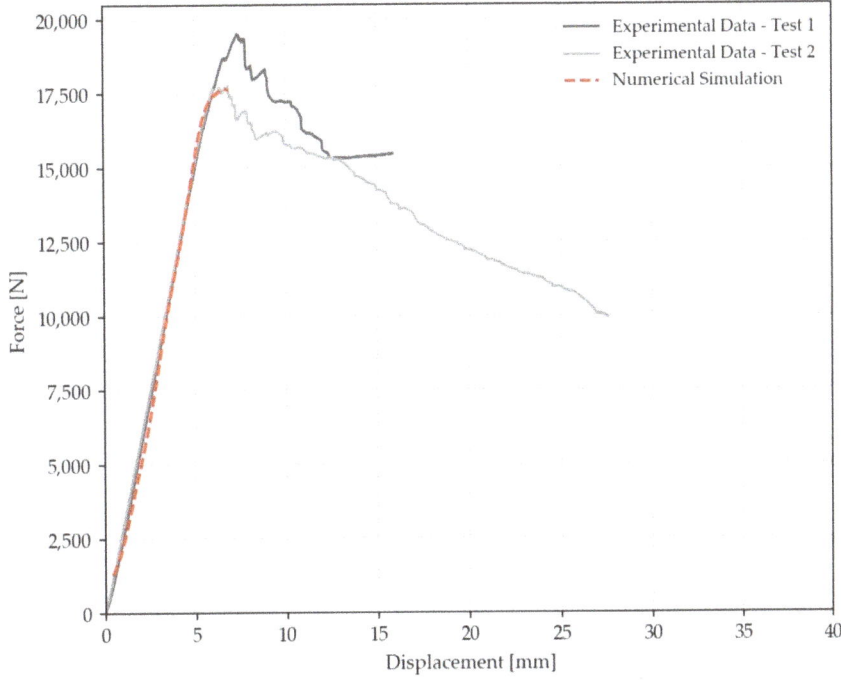

Figure 3. Experimental and numerical results of the four-point bending tests.

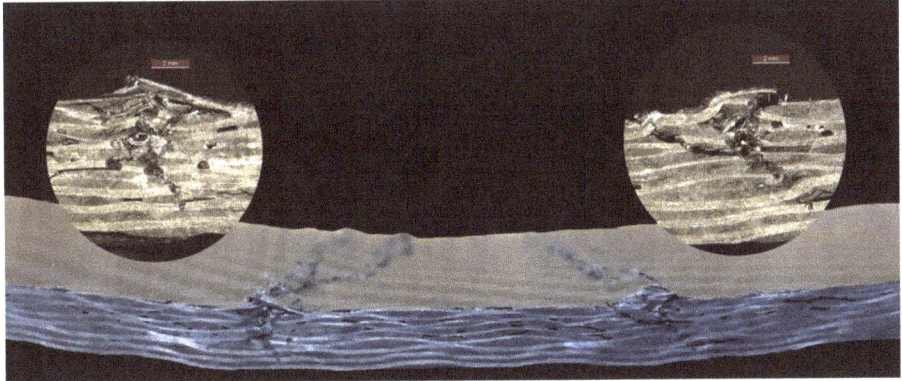

Figure 4. Magnification of intralaminar cracks: failure of the upper four layers.

3.2. Progressive Damaging and Assessment of Residual Elastic Properties through the DI_d Technique

An undamaged composite is subjected to repeated four-point bending tests. Load is increased until incipient failure occurs and then decreased. Achieved failure is not complete, as the load-carrying capacity of the composite laminate is still significant. In total, five repetitions are performed, as shown in Figure 5, where the resulting force-displacement curves are numbered as Test n.1, Test n.2, Test n.3, Test n.4, and Test n.5. As the damage increases, the plate stiffness progressively decreases. Loading and unloading cycles are performed until the residual stiffness of the composite plate is about 40% of the original value. This is shown with the Test n.6 (damaged specimen curve), which is performed only in the elastic field. A significant damage level is thus induced through the five tests.

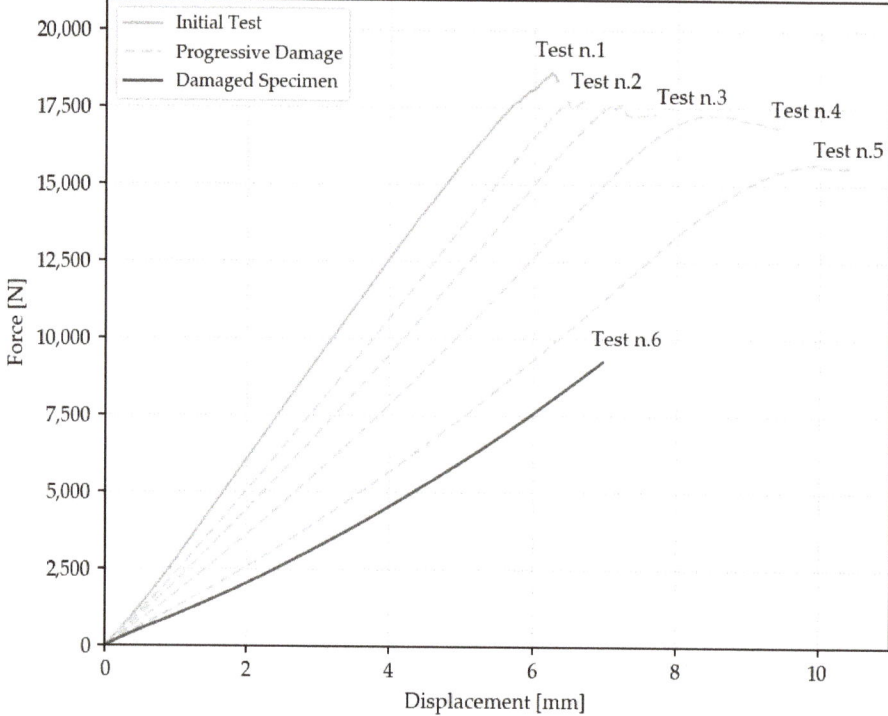

Figure 5. Progressive damaging of composite plate through four-point bending tests.

The DI_d technique is then adopted to estimate the residual elastic properties. Preliminary characterization tests allowed to evaluate the threshold energy ε_{th}, which is in this case equal to 5 J. Details of the preliminary tests are not discussed here. A complete description of the procedure can be found in [15].

Figure 6 shows the resulting correlation between the residual elastic properties of the analyzed composite and the DI_d parameter. The dotted line is thus adopted to evaluate the residual elastic properties in the progressively damaged plate. Cross markers correspond to the preliminary characterization tests.

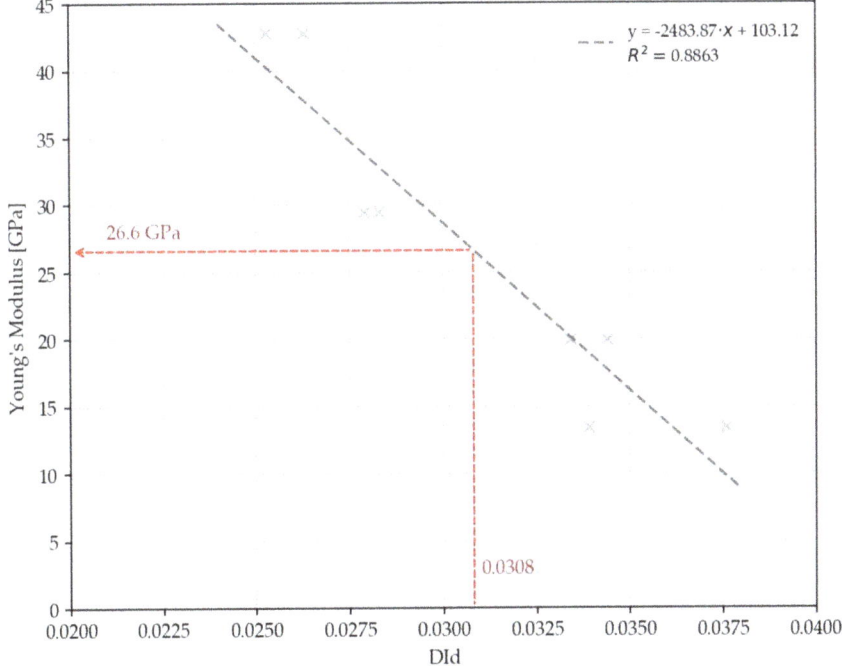

Figure 6. Residual Young's modulus evaluation with DI_d.

As reported in Figure 2, the elastic properties have been estimated at seven locations along the middle line of the composite plate. Table 2 reports the measured DI_d values and the corresponding Young's moduli, as well as the locations in x direction of the impacts.

Table 2. Residual Young's moduli and corresponding DI_d values at specific locations.

X-Coordinate [mm]	DI_d	E_{res} [GPa]
50	0.0283	32.8
75	0.0297	29.5
100	0.036	13.7
125	0.0308	26.6
150	0.036	13.7
175	0.0297	29.5
200	0.0283	32.8

In correspondence of the external noses, whereas the bending moment is null in the four-point bending test and no damage should be present, the elastic modulus is equal to 32.8 GPa, which is significantly different from the original value of 54 GPa. This can be justified by taking into account that the circular unclamped region considered for the DI_d test has a diameter of 76 mm. Consequently, elastic properties are assessed as an average of the unclamped region.

Further, as clamping boundary conditions have to be realized all around the laminate, the diameter of 76 mm limits the region where the residual properties can be measured through the DI_d. Moving outwards from the external noses, only few measurements of the residual properties can be realized and results would be affected by the damaged portion of the plate. On the basis of these considerations, we can assume that, at a distance of 38 mm from the external noses (half of the diameter), the damaged portion of the plate would not affect an ideal measurement through the DI_d. The Young's modulus is thus here considered equal to the original value of 54 GPa.

From the seven measurements of the Young's modulus and considering the two assumed external values, eight third-order polynomial curves can be constructed, according to the approach described in Section 2.2. As a consequence, a continuous variation of the Young's modulus with respect to the x-coordinate is obtained. This result is shown in Figure 7, where measured Young's moduli are marked with red squares and assumed values with black thin diamonds. The continuous curve is adopted in the numerical model to account for the variation of the Young's modulus. Layers of shells have been longitudinally divided into parts, which are represented with different colors in Figure 8. The continuous curve allows to evaluate the Young's modulus in correspondence of the x-coordinate of each part. A material card has been defined for each part with the corresponding elastic properties. Cross markers of Figure 7 represent the Young's moduli assigned to each material card.

Moving outwards from the external noses, the elastic modulus increases from 32.8 to 54 GPa and then is assumed constant. It should be noted that, as the bending moment is here null, the mechanical behavior of the numerical model will not be affected by the value of the elastic modulus estimated for this region.

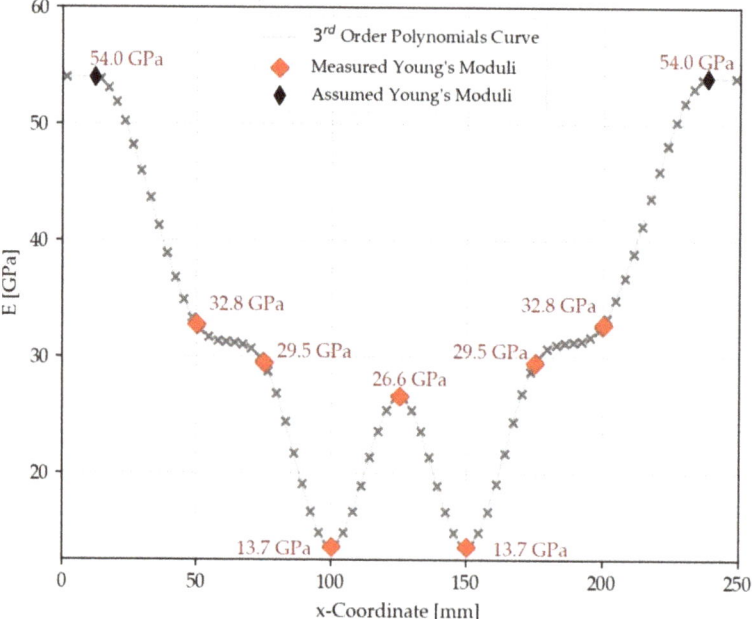

Figure 7. Variation of residual Young's modulus with respect to the x-coordinate.

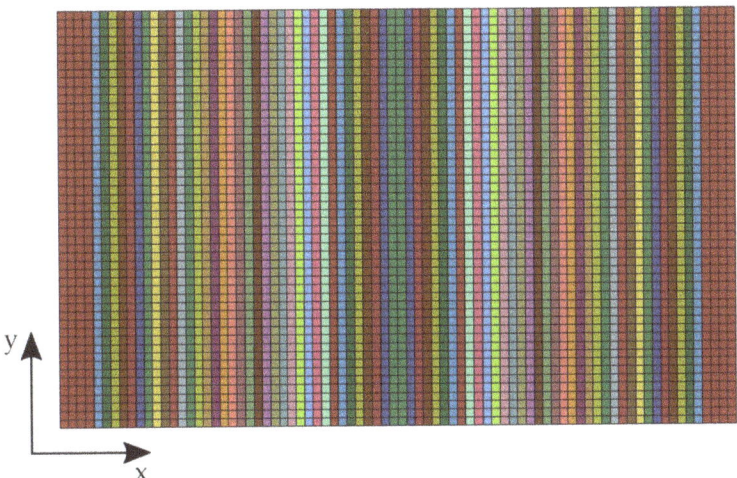

Figure 8. Top view of the finite element model of the composite laminate: longitudinal division of layers.

3.3. Validation: Comparison of Experimental and Numerical Results

The model of the composite plate with variable elastic modulus is adopted for simulating the four-point bending test. Results of the numerical force-displacement curve are then compared to those obtained with the progressively damaged plate. As shown in Figure 9, experimental and numerical results are in very good agreement, with limited discrepancies in slope. Results of numerical analysis are once again reported from 1 mm of displacement. The numerical displacement is measured on the rigid cylinders and these are initially distanced from the composite laminate, in order to avoid penetrations and numerical instabilities. Consequently, the initial path of the numerical force-displacement curve concerns the contact between laminate and cylinders and is not significant for the purpose.

Experimental response shows an increasing stiffness as the plate is bent. This hardening effect can be explained by taking into account that the considered material fails under compressive loads, as shown in Figure 4. The four-point bending tests are repeated until incipient failures occur. Consequently, cracks propagate in the upper layers subjected to compressive loads. When testing the composite plate, these cracks are progressively closed under compression. This allows to sustain loads in the failed layers and an increasing stiffness is thus obtained. The hardening effect can be also appreciated in the force-displacement curves with progressive damage of Figure 5 (dotted lines).

The limited discrepancy between experimental and numerical results highlights the accuracy of the DI_d technique in mapping the residual elastic properties of the damaged plate. Further, this result validates the proposed approach which accounts for the variable elastic modulus with a continuous polynomial curve in the finite element model.

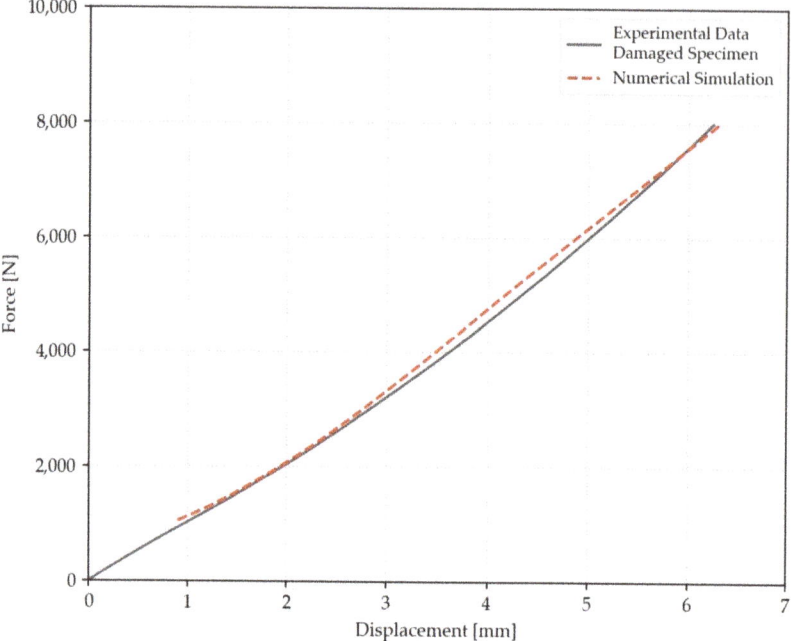

Figure 9. Comparison of experimental and numerical results of damaged specimen.

4. Conclusions

The use of the DI_d technique in combination with the finite element method was presented. A four-point bending test was performed on an 8 layers twill fabric composite laminate until complete failure occurred. Experimental results were used to set up a finite element model of the composite plate subjected to bending test.

Then, repeated four-point bending tests allowed to progressively damage an undamaged plate of the same composite material. The DI_d technique was adopted to map the residual elastic properties on the damaged plate. Seven locations have been considered along the middle line of the composite laminate. In particular, damage was mainly localized in correspondence of the inner noses. Further, it has been found that elastic properties were assessed as an average of the circular unclamped region considered for the DI_d test, whose diameter is 76 mm. The diameter of the analyzed region may thus be reduced in order to achieve very local assessments.

A continuous polynomial curve was then considered to account for the variable elastic modulus in the finite element model. The resulting force-displacement curve of the numerical analysis was in very good agreement with experimental data of damaged plate.

Therefore, the DI_d methodology allows to locally assess the residual elastic properties of damaged composite materials. By mapping the elastic properties on the component and considering the assessed values in a finite element model, a precise description of the mechanical behavior of the composite plate is obtained.

Consequently, thanks to the proposed methodology, the health state of a damaged component can be quantitatively evaluated and decisions on its maintenance can be made by defining limits on the acceptable damage level.

Author Contributions: Conceptualization, G.B.and D.P.; Data curation, C.N. and A.T.; Funding acquisition, C.N., A.T. and R.C.; Investigation, C.N., A.T. and R.C.; Methodology, G.B. and D.P.; Project administration, G.B. and D.P.; Software, C.N.; Supervision, G.B. and D.P.; Validation, C.N. and A.T.; Visualization, C.N.; Writing—original draft, C.N.; Writing—review and editing, A.T., R.C., G.B. and D.P.

Funding: This research received no external funding.

Conflicts of Interest: The authors declare no conflict of interest.

References

1. Heslehurst, R.B. *Defects and Damage in Composite Materials and Structures*; CRC Press: Boca Raton, FL, USA, 2017.
2. Ibrahim, M.E. Nondestructive testing and structural health monitoring of marine composite structures. In *Marine Applications of Advanced Fibre-Reinforced Composites*; Woodhead Publishing: Cambridge, UK, 2016; pp. 147–183.
3. Hubschen, G.; Altpeter, I.; Tschuncky, R.; Herrmann, H. *Materials Characterization Using Nondestructive Evaluation (NDE) Methods*; Woodhead Publishing: Cambridge, UK, 2016.
4. Katunin, A.; Danczak, M.; Kostka, P. Automated identification and classification of internal defect in composite structures using computed tomography and 3D wavelet analysis. *Arch. Civ. Mech. Eng.* **2015**, *15*, 436–448. [CrossRef]
5. Junyan, L.; Liqiang, L.; Yang, W. Experimental study on active infrared thermography as a NDI tool for carbon–carbon composites. *Compos. Part B Eng.* **2013**, *45*, 47–138. [CrossRef]
6. Tam, J.H.; Ong, Z.C.; Ismail, Z.; Ang, B.C.; Khoo, S.Y. Identification of material properties of composites materials using nondestructive vibrational evaluation approaches: A review. *Mech. Adv. Mater. Struct.* **2017**, *24*, 971–986. [CrossRef]
7. ASTM. *E1876-15 Standard Test Method for Dynamic Young's Modulus, Shear Modulus, and Poisson's Ratio by Impulse Excitation of Vibration*; ASTM International: West Conshohocken, PA, USA, 2015.
8. ASTM. *C1259-15 Standard Test Method for Dynamic Young's Modulus, Shear Modulus, and Poisson's Ratio for Advanced Ceramics by Impulse Excitation of Vibration*; ASTM International: West Conshohocken, PA, USA, 2015.
9. Paolino, D.S.; Geng, H.; Scattina, A.; Tridello, A.; Cavatorta, M.P.; Belingardi, G. Damaged composites laminates: Assessment of residual Young's modulus through the Impulse Excitation Technique. *Compos. Part B Eng.* **2017**, *128*, 76–82. [CrossRef]
10. Garnier, C.; Pastor, M.L.; Eyma, F.; Lorrain, B. The detection of aeronautical defects in situ on composite structures using Non-Destructive Testing. *Compos. Struct.* **2011**, *93*, 36–1328. [CrossRef]
11. Belingardi, G.; Cavatorta, M.P.; Paolino, D.S. A new damage index to monitor the range of the penetration process in thick laminates. *Compos. Sci. Technol.* **2008**, *68*, 2646–2652. [CrossRef]
12. Belingardi, G.; Cavatorta, M.P.; Paolino, D.S. Repeated impact response of hand layup and vacuum infusion thick glass reinforced laminates. *Int. J. Impact Eng.* **2008**, *35*, 609–619. [CrossRef]
13. Belingardi, G.; Cavatorta, M.P.; Paolino, D.S. On the rate of growth and extent of the steady damage accumulation phase in repeated impact tests. *Compos. Sci. Technol.* **2009**, *69*, 1693–1698. [CrossRef]
14. Belingardi, G.; Cavatorta, M.P.; Paolino, D.S. Single and repeated impact tests on fiber composite laminates: Damage index vs. residual flexural properties. In Proceedings of the 17th International Conference for Composite Materials. Edinburgh, UK, 27–31 July 2009.
15. Tridello, A.; D'Andrea, A.; Paolino, D.S.; Belingardi, G. A novel methodology for the assessment of the residual elastic properties in damaged composite components. *Compos. Struct.* **2017**, *161*, 435–440. [CrossRef]
16. Yam, L.H.; Wei, Z.; Cheng, L. Nondestructive detection of internal delamination by vibration-based method for composite plates. *J. Compos. Mater.* **2004**, *38*, 98–2183. [CrossRef]
17. LSTC. LS-DYNA Keyword Manual Volume I. 2019.
18. LSTC. LS-DYNA Keyword Manual Volume II. 2019.
19. Muflahi, S.A.; Mohamed, G.; Hallett, S.R. Investigation of delamination modeling capabilities for thin composite structures in LS-DYNA. In Proceedings of the 13th International LS-DYNA Users Conference, Detroit, MI, USA, 8–10 June 2014.

20. Williams, K.V.; Vaziri, R.; Floyd, A.M.; Poursatip, A. Simulation of damage progression in laminated composite plates using LS-DYNA. In Proceedings of the 5th International LS-DYNA Users Conference, Southfield, Dearborn, MI, USA, 9–11 April 1998.
21. Moncayo, E.D.; Wagner, H.; Dreschler, K. Benchmarks for composite delamination using LS-DYNA 971: Low velocity impact. In Proceedings of the German LS-DYNA Forum, Frankenthal, Gemany, 11–12 October 2007.
22. ASTM. *D6272–17. Standard Test Method for Flexural Properties of Unreinforced and Reinforced Plastics and Electrical Insulating Materials by Four-Point Bending*; ASTM International: West Conshohocken, PA, USA, 2017.
23. ASTM. *D5628-10. Standard Test Method for Impact Resistance of Flat, Rigid Plastic Specimens by Means of a Falling Dart (Tup or Falling Mass)*; ASTM International: West Conshohocken, PA, USA, 2010.
24. LSTC. LS-DYNA Theory Manual. 2019.

© 2019 by the authors. Licensee MDPI, Basel, Switzerland. This article is an open access article distributed under the terms and conditions of the Creative Commons Attribution (CC BY) license (http://creativecommons.org/licenses/by/4.0/).

Review

A Review of Non-Destructive Damage Detection Methods for Steel Wire Ropes

Ping Zhou [1], Gongbo Zhou [1,*], Zhencai Zhu [1], Zhenzhi He [2], Xin Ding [1] and Chaoquan Tang [1]

1. Jiangsu Key Laboratory of Mine Mechanical and Electrical Equipment, School of Mechatronic Engineering, China University of Mining and Technology, Xuzhou 221116, China
2. School of Mechanical and Electrical Engineering, Jiangsu Normal University, Xuzhou 221116, China
* Correspondence: gbzhou@cumt.edu.cn; Tel.: +86-182-0520-7100

Received: 22 May 2019; Accepted: 5 July 2019; Published: 9 July 2019

Abstract: As an important load-bearing component, steel wire ropes (WRs) are widely used in complex systems such as mine hoists, cranes, ropeways, elevators, oil rigs, and cable-stayed bridges. Non-destructive damage detection for WRs is an important way to assess damage states to guarantee WR's reliability and safety. With intelligent sensors, signal processing, and pattern recognition technology developing rapidly, this field has made great progress. However, there is a lack of a systematic review on technologies or methods introduced and employed, as well as research summaries and prospects in recent years. In order to bridge this gap, and to promote the development of non-destructive detection technology for WRs, we present an overview of non-destructive damage detection research of WRs and discuss the core issues on this topic in this paper. First, the WRs' damage type is introduced, and its causes are explained. Then, we summarize several main non-destructive detection methods for WRs, including electromagnetic detection method, optical detection method, ultrasonic guided wave detection method, and acoustic emission detection method. Finally, a prospect is put forward. Based on the review of papers, we provide insight about the future of the non-destructive damage detection methods for steel WRs to a certain extent.

Keywords: non-destructive damage detection; steel wire ropes; review; electromagnetic detection; optical detection; ultrasonic guided wave

1. Introduction

Steel wire rope was invented in 1834. It has many advantages, such as high tensile strength, stable and reliable operation, and a strong capacity for dynamic load and overload. It is widely used in lifting, transportation, and traction equipment/systems. The main equipment/systems include mine hoists, ropeways, cranes, oil rigs, elevators, and cable-stayed bridges [1–4]. During the service of steel wire ropes (WRs), defects and damages like wire breakage, wear, rust, fatigue, strand breakage, and even sudden breaking will inevitably occur due to various reasons. Therefore, academia and industry have been trying to explore various methods to detect damages of WRs to guarantee its reliability and safety [5].

However, due to the complexity of WR structure, the diversity of working environment, and the limitation of detection methods, non-destructive detection of WRs has become an old and difficult problem [5]. The development of related technologies is slow, and it is difficult to achieve efficient and reliable industrial application. At present, manual inspection under low-speed operation (visual inspection combined with manual touch) is still the main method in most cases, and the electromagnetic detectors are used in a few cases, which cause problems such as missed detection, false detection, and failure to detect in time, resulting in frequent occurrence of faults and accidents [6]. Taking the mining industry as an example, on 29 December 2010, a wire rope at a resort in Maine broke and more than 200 people in the cable car were trapped in the air and 9 people were injured; On 9 March 2017, a wire rope of a multi-rope friction mine hoist of the Mining Company of Heilongjiang Longmei Group suddenly broke during the lifting process causing a cage fall (cable fire aggravates wire rope breakage). As a result, 17 people trapped in the cage were killed. Therefore, it is of great significance to study reliable and efficient non-destructive detection methods for WRs.

There are many methods for WRs damage detection [7], including electromagnetic detection, optical detection, ultrasonic guided wave detection, acoustic emission detection, ray detection, eddy current detection, and vibration detection. The most important methods are electromagnetic detection and optical detection (machine vision method). In particular, the instruments based on electromagnetic detection have been gradually applied in the market. Although the optical detection method can intuitively grasp the surface morphology characteristics of WRs, it is seldom used because of the influence of algorithm performance and the inability to detect internal defects. In addition, other methods are still in the theoretical and laboratory stage for many reasons. Non-destructive detection of WRs is an important research field, and there are some literature reports on various methods. However, there is still a lack of a reasonable and systematic summary to illustrate the current research situation and point out the possible research directions in the future.

Based on this, we first summarize the research status of each important detection method and analyze its shortcomings, then we summarize the existing methods and put forward the prospects. To our best knowledge, this article will make up for the lack of literature review in the field of non-destructive detection of WRs, which can bridge this gap and promote the development of non-destructive detection technology of WRs. The contributions of this paper can be summarized as follows: (1) The literatures in the field of non-destructive testing of wire rope are summarized in detail, including electromagnetic detection method, optical detection method, ultrasonic guided wave method, and acoustic emission method. (2) Shortcomings of these methods are pointed out, and comparisons are made among them. (3) Each important detection method is prospected in combination with the current technological situation, which is expected to promote the development of this direction.

The work is arranged as follows: In Section 2 the WRs damage type is introduced and its causes are explained. In Section 3 several main WRs non-destructive detection methods are reviewed. Their main disadvantages are summarized in Section 4. The prospect is presented in Section 5.

2. Damage Types of Wire Ropes

Steel wire ropes will be damaged in various forms when used, which will reduce the strength of the WRs and pose a potential threat to the system safety. Various damage conditions and distribution have different effects on the strength reduction of WRs, and the whole WR is often scrapped due to a certain section of the WR. Therefore, the study of various damages of WRs will be beneficial to the correct evaluation of its state [8,9]. The damage types of WRs include wire breakage, wear, deformation, rust, and fatigue. Among them, fatigue has a variety of representations, such as internal and external cracks, internal and external wire breakages, and slack. The common types are shown in Figure 1.

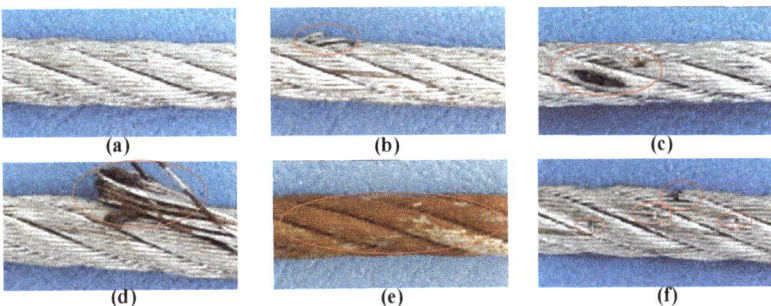

Figure 1. Sample pictures of wire rope damage. (**a**) Health; (**b**) wire breakage; (**c**) wear; (**d**) deformation (broken strand); (**e**) rust; and (**f**) fatigue (fatigue breaking).

In practice, the above damage types will interact with each other, such as rust will aggravate the wear process, wear will promote wire breakage, fatigue will promote the generation or aggravate the evolution of wire breakage and wear, and the comprehensive impact of wear, wire breakage, fatigue, and rust will promote broken strands and aggravate the scrap of wire rope. When minor faults evolve to a certain extent or suffer sudden impact load, strand breakage will occur, which will obviously affect the life of WR. Because of the different working conditions and environment, the development speed and degree of damages are also different. In current standards, the residual strength is generally obtained by finding the defects of WR in service to infer the strength loss, then, whether the WR is scrapped is judged according to the safety factor of WR and the rules of use [9].

The description of different damage types of WRs is shown in Table 1.

Table 1. Types and description of wire rope (WR) damage.

Types	Description
Wire breakage	Fatigue breakage and wear breakage are the main causes of wire breakage. Wire breakage will reduce the WR strength and increase the potential safety hazards of WRs.
Wear	Wear is one of the most common phenomena of WR, which usually evolves from normal wear to failure wear. The wire wear will reduce the tension it can bear.
Rust	Rust is a fault phenomenon of WRs caused by the chemical and electrochemical action of surrounding medium, which has a significant impact on the life of WRs.
Deformation	Impact on the WR causes damages (including flattening, strand relaxation, kinking, bending, and strand breakage) to the original structure, which accelerates the scrap.
Fatigue	After repeated bending for a certain number of times or repeated stretching and twisting, fatigue occurs, which leads to changes in the internal and external performance of the WR.

In general, in the WR damage detection and diagnosis practice, according to the characteristics of WR damages, the WR damages can be divided into two categories [10]: local flaw damage type (LF) and loss of metallic cross-sectional area type (LMA). LF type refers to the damage locally generated in the WR, mainly including internal and external breakage, corrosion pit, and deep wear of steel wire, characterized by a sudden decrease in the metallic cross-sectional area of the WR; LMA type refers to the slow reduction of the effective metallic cross-sectional area in the long range along the axial direction of the WR, which mainly includes wear, long-distance rust, and rope diameter reduction.

3. Non-Destructive Detection Method for Steel Wire Ropes

The following is a literature review of electromagnetic detection, optical detection, ultrasonic guided wave, acoustic emission detection method, and other detection methods.

3.1. Electromagnetic Detection Method

Electromagnetic detection is one of the oldest methods in non-destructive testing field and has been applied earliest to WRs detection. In this detection method, there are two different excitation modes including coil excitation and permanent magnet excitation, and the commonly used sensor types are the Hall sensor and the induction coil.

The detection principle diagram is shown in Figure 2, the main principle is: after the WR is excited by coil or permanent magnet, the surface and internal damages will change the magnetic flux in the WR. Then the signal is detected by the induction coil or integrated sensor (e.g., Hall sensor and magnetoresistance sensor), and the type or degree of damages can be obtained to a certain extent through further processing and analysis of the signal [8–10]. Figure 2a is the main magnetic flux detection method based on coil excitation, which adopts the coaxial magnetization method, namely, taking the measured wire rope as the core, and the wire rope segment between the two coils is magnetized by the coil with current. Figure 2b is the main magnetic flux detection method based on permanent magnet excitation, in which the method of inter-pole magnetization is adopted, taking the wire rope as a component of the magnetic circuit, and the induction coil is used to detect the change of magnetic flux inside the wire rope caused by damage. Figure 2c is a method of return magnetic flux detection based on permanent magnet excitation, whose excitation mode is the same as Figure 2b, but the Hall sensor is used to detect the flux density in the magnetic circuit formed by excitation device, air gap and wire rope. Figure 2d is leakage magnetic flux detection method based on permanent magnet excitation, whose excitation mode is the same as Figure 2b,c, however, the difference is that the sensor is used to collect the leakage magnetic field information caused by the damage.

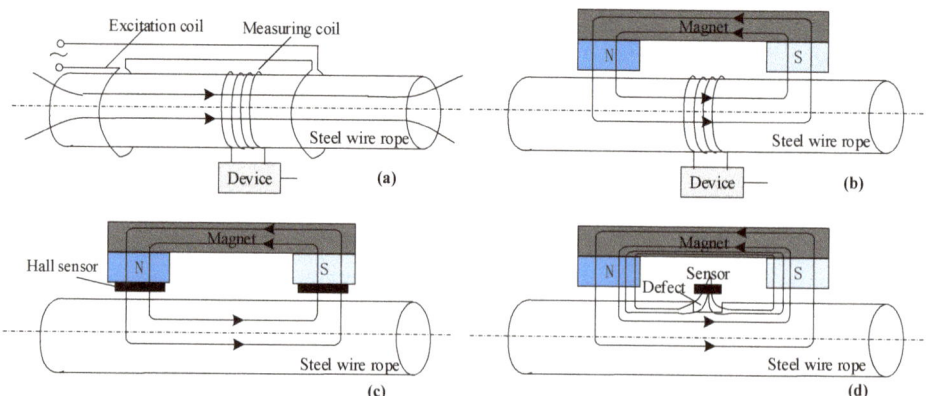

Figure 2. Electromagnetic non-destructive detection principle of steel wire rope. (**a**) Main magnetic flux detection method based on coil excitation; (**b**) main magnetic flux detection method based on permanent magnet excitation; (**c**) return magnetic flux detection method based on permanent magnet excitation; and (**d**) leakage magnetic flux detection method based on permanent magnet excitation.

In 1906, South Africa developed the first non-destructive testing device for WRs based on electromagnetic principle. Over the past 100 years, electromagnetic detection technology has made great progress [11]. The research of electromagnetic detection technology mainly includes: the mechanism of electromagnetic detection signal under the influence of defect parameters (defect width, depth, number of broken wires, etc.) or the theoretical model of the relationship among them; the magnetization method and device of WRs; the electromagnetic signal detection means such as the coil sensor, Hall sensor, and magnetoresistance sensor. It is noted that the signal processing and recognition methods are accompanied with the above three aspects. Following is a literature review of the above three aspects.

3.1.1. Mechanism and Model

For electromagnetic detection mechanism and theoretical model, many studies have been carried out by researchers at home and abroad.

For instance, Norouzi et al. [12] used the finite element analysis method to optimize the pole shoe length, and the optimized results help make the magnetic flux distribution in the measured parts as uniform as possible, which was more conducive to magnetic leakage detection. In [13], Zhao et al. performed finite element analysis for the magnetic flux leakage (MFL) distribution of WR typical defects based on strong magnetic detection method and obtained the relationship between MLF signal and detection distance, damage depth, damage axle width and internal wire breakage. In [14], Lenard et al. paid attention to oil and gas pipelines, and analyzed the influence of line pressure, bending and residual stresses on MFL signal in MFL detection process through experiment and finite element simulation. To study the effects of probes and probe lift-off on the defect leakage magnetic field, Krzywosz [15] and Kalwa et al. [16] respectively carried out research and proposed that the smaller the lift-off value of magnetic sensitive probes, the better.

In order to apply the magnetic dipole to the three-dimensional magnetic field modeling of MFL, Dutta et al. [17] first presented a three-dimensional (3D) magnetic leakage model of ferromagnetic materials surface fracture based on magnetic charge theory. Based on [17], an improved 3D MFL model for numerical analysis is proposed by Trevino et al. [18], which can be used to detect surface fracture of ferromagnetic materials under saturated magnetization. To further explain the formation principle of MFL, Sun et al. [19] explained the mechanism of MFL in engineering from the perspective of magnetic refraction, and found that the signal component in defect detection was different from that in traditional MFL analysis, which was produced by protruding defects. Considering the effect of tensile stress on the strength of the MLF of a steel WR defect, Gao et al. [20] loaded a variable force onto reproduced samples of typical defective WRs relying on an inspection platform. The experimental results showed that the peak-to-peak values of the MFL signal from all defective WRs increased with tensile stress, in an approximately linear relationship.

3.1.2. Magnetization Method

The magnetizer is an important component of WR non-destructive detection system. Its excitation structure and mode will produce different magnetization effects [21]. The commonly-used magnetization methods mainly include the coil electromagnetic magnetization method and permanent magnetization method.

Thus far, Kang et al. [22] analyzed the different WR magnetization effect of different magnetization methods by finite element method and obtained the optimum size of feed-through magnetizer. In [23], the equipment was operated according to the principle of strong magnetic detection, which realized magnetization saturation by magnetizing the wires with a pre-magnetic head. Jomdecha et al. [24] improved the traditional current excitation device, so that the excitation intensity of the device can be controlled by adjusting the excitation power supply and the coil. Wang et al. [25] designed an excitation device that can restrain the fluctuation of lift-off distance and can improve the structure of the detection device. Song et al. [26] studied whether AC magnetization in U-yoke would cause eddy current disturbance field. The results showed that the axial component of MFL of crack was independent of excitation frequency and intensity.

In the practical application, for thin WR in strong electromagnetic interference environment, Yan et al. [27] proposed an electromagnetic non-destructive testing method. In this method, a simplified magnetic circuit was proposed to excite WR, and the defect detection of thin WR in electromagnetic interference environment was realized. A new WR tester was established based on MLF principle in [28], and each arc segment subtends an angle of 22.5 degrees at the center by cutting two rings of NdFeB along axial direction into 32 equal arc segments, which were then parallelly magnetized in magnetizer. To detect WR damage in coal mine, Sun et al. [29] proposed an opening electric magnetizer through the magnetic control for a C-like electric loop-coil, which has an opening structure capable of directly encompassing and centering the endless object in it. The magnetizer can solve problems, i.e., installation inconvenience and on-line detection difficulty in the non-destructive testing of coal mine wire rope fixed at both ends.

3.1.3. Detection Sensor

(1) Coil sensor

Coil sensor is widely used because of its low cost and ease of use. In recent years, its application in WRs non-destructive detection is still further studied.

For example, Liu et al. [30] assessed the surface and internal flaws using a new type of sensor to measure the biased pulse magnetic response in a large-diameter steel stay cable. Two parallel connected flexible flat coils fed with a biased pulse current were deployed by the sensor as the electromagnet for cable magnetization. Yan et al. [31] put forward a kind of iron core as coil winding skeleton for the wire rope non-destructive testing relying on the theoretical analysis and 3D transient magnetic field simulation. The experiment results proved that the signal to noise ratio of coil with the iron core proposed in this paper increased almost six times, making it easier for defect analysis. Aiming at detecting the change of surface leakage magnetic field, Fedorko et al. [32] designed a pair of induction coils of different sizes and analyzed the static magnetic field distribution based on finite element simulation. Sun et al. [33] designed a new MLF sensor based on open magnetizing method. Experimental comparisons between the open and yoke probes for on-line automated monitoring were conducted, which confirmed the characters of smaller magnetic interaction force, less wear, and damage in this method were in contrast to the traditional on-line automated structural health monitoring technology.

(2) Integrated sensor

Through using integrated sensors to detect MLF of the WR surface, the circumferential and axial locations of defects can be founded, so as to analyze whether the defects belong to centralized or decentralized wire breakage, and to make a further accurate judgment on the damage of WRs [11].

In the work of [34–38], the Hall sensors were used. Xu et al. [34] designed a sensor module based on Hall sensor to test the leakage magnetic field of stay cables online and compared three different filtering algorithms to obtain the leakage magnetic field of artificial damage inside the cable. Wang et al. [35] proposed a new method to detect magnetic excitation in WRs by combining the structural models for dynamic magnetic field balancing and magnetic focusing. A Hall-element array and the magnetic focusing technique were used to reduce the interference produced by the interactions between the environmental magnetic fields and the wire rope strand. A MLF method is adopted by Kim et al. [36] for the detection of local damage when inspecting WRs, and a multi-channel MFL sensor head was fabricated by adapting to the wire rope with a Hall sensor array and magnetic yokes, the accuracy and reliability were evaluated based on the comparison of the repeatedly estimated damage size and the actual damage size. Tian et al. [37] designed an optimal model for magnetic excitation to develop a non-destructive sensor using a Hall-effect sensor for coal mine hoist wire ropes. Then, a new detection method called the permanent magnet co-directional excitation flux-weakening method was put forward in [38]. The emulation and experimental results revealed that the defect detection sensor proposed in the paper can improve the magnetic signal from the defect by six to eight times.

In addition, the magnetoresistance sensors were used in the work of [39,40]. Singh et al. [39] proposed giant magnetoresistive-based MLF technique for condition monitoring of 64 mm-diameter steel track rope which can be used to transport coal. Two saddle coils were used to magnetize the rope in this technique, and a giant magnetoresistive sensor was able to detect the tangential component of leakage magnetic flux from flaws. In addition, the technique can detect 2 mm-deep flaws with good signal-to-noise ratio and solve the flaw whose interval is more than 3.2 mm. Tunnel magnetoresistive (TMR) devices was first employed to form a circular MLF sensor by Liu [40], aiming to detect slight WR flaw. Two versions of this tailor-made circular TMR-based sensor array were put forward to inspect WRs with the diameters of 14 mm and 40 mm, respectively.

3.2. Optical Detection Method

Optical detection method is an efficient non-destructive detection method, which has been developed in the field of WR surface damage detection. It can intuitively grasp the damage situation of WR surface. At the same time, with the development of WR oil pollution removal method/device [41], the influence of oil pollution gradually decreases.

The principle of optical detection method is shown in Figure 3, which generally includes two parts: image acquisition and damage diagnosis. The image acquisition generally uses high-speed camera to collect the surface image of WRs, and the damage diagnosis includes image preprocessing and pattern recognition. WR damage detection based on pattern recognition is a potential research direction. Specifically, the hardware and software of the image acquisition system are configured according to the actual situation; the collected images are first processed by pretreatment method like cutting or filtering. Then, feature extraction methods (such as local binary pattern and gray level co-occurrence matrix) are used to extract features, and next dimensionality reduction is carried out on the established feature datasets. Finally, machine learning classifiers (such as support vector machines, artificial neural networks, etc.) are used for training and testing to obtain robust classifiers for state recognition of unknown images [42].

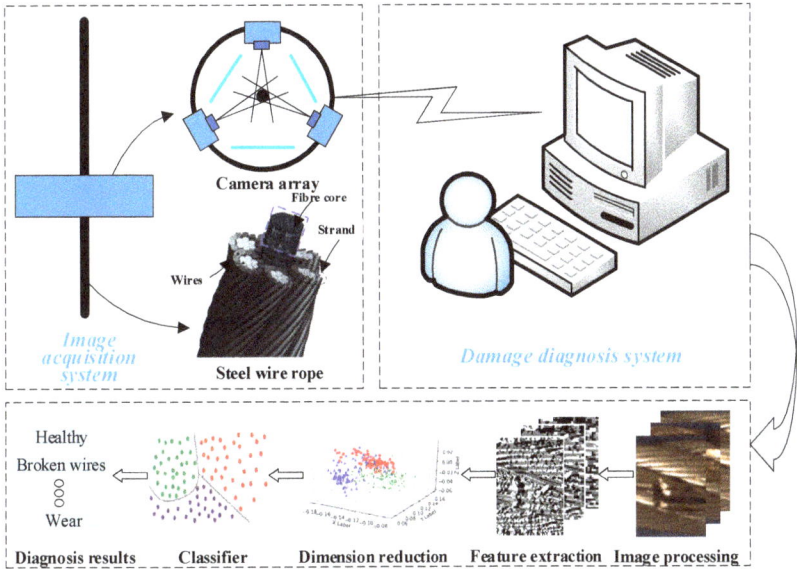

Figure 3. Optical detection principle for steel wire rope.

Based on camera and offline processing algorithm, Vallan et al. [43] established a measurement system which measured the change of rope length according to the characteristics of rope twist to detect the damage of rope. For elevator rope, Yaman et al. [44] proposed a fault monitoring method based on image processing and autocorrelation analysis, which realized the defect detection of steel rope to a certain extent. In [45], Sun et al. realized the identification and segmentation of WRs boundaries through image processing and clustering methods. Platzer et al. [46] proposed using Hidden Markov field model to locate the defects of WR, the results demonstrated that the detection performance of this method is better than that of the previous time invariant system classification method. Then, they compared the performance of different texture features in the detection of WR surface defects in [47]. For the defects of WR, the histogram of oriented gradient feature was the best, followed by the local binary patterns feature. In the work of [48,49], Wacker et al. established a probabilistic appearance model as a representation of normal surface changes combined with the structure and appearance of the WR and realized the detection of WR abnormal surface. For surface defects of thin metallic wires, an automatic optical detection technique was presented by Sanchez-Brea et al. in [50]. This technique was based on the intensity variations on the scattered cone generated when the wire is illuminated with a beam at oblique incidence.

3.3. Ultrasound Guided Wave Method

Ultrasonic guided waves method (UGW) is one of the promising methods for investigation of non-homogeneities for the internal structure of WRs [51]. The measurement schematic diagram is illustrated in Figure 4, which is consisted of the computer, the ultrasonic measurement system, the preamplifier, the transmitter, and the receiver, detailed principles are available for reference [52]. The wave is launched on WRs and propagates along the strand of WRs. Acoustic reflection occurs when the sound wave passes through the damage. After collecting and analyzing the echo at the receiving end, the damage inside and outside the WR can be obtained [53].

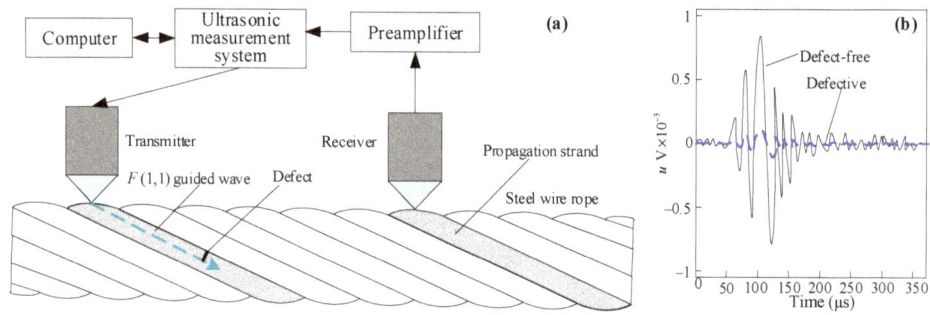

Figure 4. Non-destructive testing principle of steel wire rope by ultrasonic guided wave method [52]. (a) Detection system schematic diagram; (b) sketch map for waveforms of defect-free and defective signal.

For purpose of the generation and reception of ultrasonic longitudinal guided waves in seven-wire steel strands in a pitch catch arrangement, Liu et al. [53] optimized the magnetostrictive transducers configuration for both transmitter and receiver. Consequently, magnetostrictive transducers with an optimized configuration, including permanent magnet distribution and multilayer coil connection, could be efficiently used for the inspection of seven-wire steel strands by using UGW in a pitch catch arrangement. In [54], Treyssede et al. studied the transmission characteristics of elastic guided wave in multi-stranded WR. The dispersion curve of spiral WR was obtained by semi-analytical finite element method and signal energy calculation, and the optimum excitation and receiving position of ultrasonic guided wave was calculated. Vanniamparambil et al. combined UGW, acoustic emission technology and digital image correlation, and fused the features obtained by the three technologies to achieve non-destructive testing [55]. Xu et al. [56] studied the effect of different frequencies of UGW on the detection rate of steel WR defects. It was found that the higher frequency of guided waves in the steel WR lead to the faster energy attenuation, and the receiving length of elastic waves increased with the frequency. The propagation characteristics of various UGW between strands and cores of WR and their effects on the penetration depth of WR were studied in [57] by Raisutis et al. In the work of [58], the efficiency of employing the magnetostriction of ferromagnetic materials were studied relying on the UGW method for WRs damage inspection, and the location and severity of damages were approximately identified and characterized using the short-time Fourier transform and wavelet analysis.

3.4. Acoustic Emission Detection Method

Acoustic emission (AE) refers to the transient elastic wave phenomena produced by the rapid release of energy in the local area of the material under the influence of the outside world [59]. Acoustic emission sensors can convert transient elastic waves into electrical signals based on piezoelectric effect. The change of internal damage can be inferred from the analysis of electrical signals [60]. The AE testing system is shown in Figure 5a (in practical application, specific device structure and detection method should be designed according to specific detection objects) [61]. The testing principle are as follows: the two ends of the WR specimen are installed on the tensile test bench, and two clamps are clamped in the middle of the WR to help install two acoustic emission sensors. The loading parameters of the test bench are controlled by a load controller. Acoustic emission equipment is used to collect and analyze the collected data in the process of loading to obtain the occurrence and location of defects on the WR. The waveform of broken wire is shown in Figure 5b. AE technology can qualitatively identify the broken wire signal, which has the characteristics of high amplitude and high absolute energy [62]. However, because this method can only be used in the static load part, the signal-to-noise ratio is low, the cost of the instrument is high, and it is difficult to measure dynamically, it is still in the laboratory stage at present, and is difficult to be effectively used in engineering.

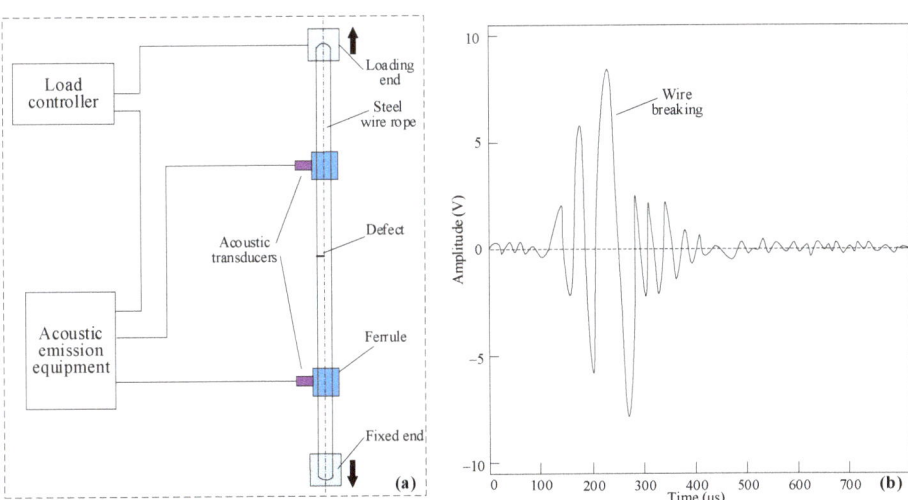

Figure 5. Acoustic emission detection for steel wire rope [61]. (**a**) Detection system schematic diagram; (**b**) sketch map for waveform of defects and wire breaking signal.

The research of relevant experimental methods and principles is active. The most systematic research on detecting broken wire of WR by AE technology was carried out by Casey et al. [63–67] in the 1980s. Based on a large number of experiments, they put forward an analysis method of amplitude distribution, that is, using the amplitude of AE signal to judge wire breakage. Due to the limitation of instrument level at that time, the result was not very ideal. Casey et al. [68] summarized the research of AE technology in the defect monitoring of WR at first, pointed out that the most important application of AE technology in the WR monitoring is the detection and location of broken wires, and discussed in detail the influence of the WR structure, the size and the number of broken wires on the AE monitoring results. Subsequently, with the development of technology, this method has been further studied. AE technology was adopted to quantitatively detect wire breakage in WR by Shao et al. [69]. The appropriate timing parameters were determined by waveform analysis, and the broken wire event was characterized as a single AE impact signal, which realized the quantitative expression of broken wire. For the problem of that inter wire fretting in ropes could cause excessive low amplitude noise, a new method based on modern signal processing techniques proposed by Ding et al. [70] could be applied to solve it. Drummond et al. [61] monitored the fatigue process of wire ropes by AE and established the relationship between the AE signal characteristics and wire breaks. Bai et al. [62] used AE technique to monitor the tensile testing process for two kinds of elevator wire ropes, and in this work AE signals from wire breaks were obtained and analyzed by AE parameters and waveforms. In the work of [71], Li et al. proposed an innovative monitoring method that used waveguides to draw out the steel wires at the end of the cable, which can effectively capture the AE signal produced by defects/wire breaking and accurately locate the transverse position of the wire breaking of the cable. Li et al. [72] developed a damage assessment and warning method for stay cables based on the AE technique and fractal theory. In this work, the fatigue test of composite cable was conducted, the fractal dimension of AE signal was analyzed, the damage index according to the fractal dimension was established.

3.5. Other Detection Methods

Only a few literature reports involve other non-destructive detection methods of WRs, such as eddy current detection [73,74] and ray detection [75,76], and are in the laboratory stage for many reasons. The detection principles are depicted in Figure 6. Figure 6a represents the eddy current method and Figure 6b explains the ray method.

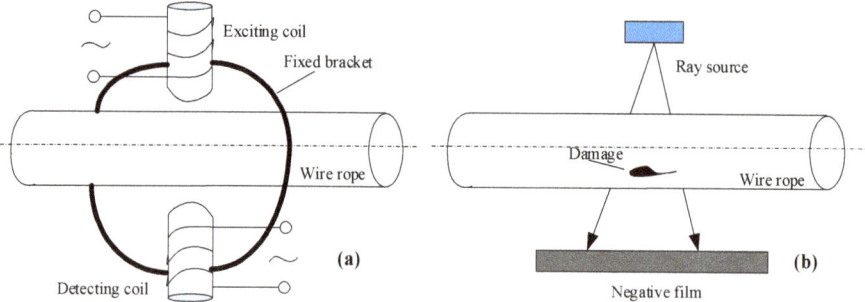

Figure 6. The other detection method principles. (**a**) Eddy current method [74]; (**b**) ray method [75].

Eddy current non-destructive testing technology takes eddy current effect produced by metal conductor in alternating magnetic field as its working principle. It has the advantages of high sensitivity, fast detection speed, non-contact, etc. [73]. Using low frequency transmission eddy current testing method, Cao et al. [73] designed an adjustable annular eddy current testing device with symmetrical probe arrangement along the radial direction to perform the quantitative detection analysis of WR breakage defects. Meanwhile, Cao et al. [74] developed an experimental eddy current sensor and computer measuring system to obtain characteristic data for rope samples made in laboratory. The characteristic data are identified by the RBF network, and the identification results show the proposed evaluation method is feasible.

When non-destructive testing of WR is carried out by ray method, the damage situation inside and outside of the WR can be clearly presented on the negative film because of the different absorption ability of radiation rays between the surface/internal defect and non-defect of the WR [75,76]. For this reason,

Zhang et al. [75] proposed a new WR detection system based on X-ray digital imaging technology, and introduced the X-ray digital imaging technology, system structure and working principle. A detection system of bridge cables using gamma ray is proposed by Peng et al. [76]. The exposure time and sensitivity of steel cables in gamma ray detection were studied. Meanwhile, the comparison between the ray detection method and the electromagnetic detection method was performed. The actual measurement demonstrated that the research results can improve the safety of WR and ensure the bridge safety assessment.

4. Existing Shortcomings

In summary, we find that each non-destructive detection method of WRs has its own advantages, but there are still many shortcomings [9,77,78], as shown in Table 2.

Table 2. Comparison of several wire rope detection methods.

Methods	Measurement Principle	Presentation Way	Advantages	Main Disadvantages
Artificial visual method	Inspect the WR surface at a slow speed	Direct analysis and judgment	Simple and direct judgment of surface damage	Time-consuming, and the result is affected by oil pollution and man-made factors
Electromagnetic detection method [7–10]	Measure leakage magnetic flux	Graph line	Qualitative determination of wire breakage, rust, wear and other defects	Difficult to make quantitative measurement and distinguish defects exist at the same time
	Measure main magnetic flux	Graph line	Measure the change of WR metal cross section area	Not suitable for detecting broken wire, especially when the fracture is not obvious
Optical detection method [42–50]	Detect WR surface using camera	Image	Intuitive, and high accuracy of detection	Influenced by environmental factors, data size and machine learning algorithm performance
Ultrasound guided wave method [51–58]	Ultrasound propagates in the medium	Echogram	Detect wire breakage, long distance of single detection	Weak anti-noise ability, cannot reflect the condition of WR in detail
Acoustic emission method [61–72]	Measure the ultrasound emitted by the structural change of steel wire rope	Sound transmission analysis	Detect wire breakage and deformation	Can only be used in the static load part, has low signal-to-noise ratio, high instrument cost and is difficult to measure dynamically.
Eddy current method [73,74]	Eddy current effect	Graph line	Detect wire breakage and rust	Skin effect affects wire breakage detection, and the signal-to-noise ratio is low
Ray method [75,76]	Radiation vertical to the rope with strong X-ray/gamma-ray	Image	Determine wire breakage	Instruments and radiation protection devices are expensive and cannot be continuously measured for long exposure periods

In view of the problems existing in electromagnetic detection, optical detection, ultrasonic guided wave method, acoustic emission method, and other detection methods, the following are specified in detail:

(1) In the field of WR electromagnetic detection, the existing theory and technology have developed in depth. However, according to the current research situation, the problems in the electromagnetic non-destructive detection of WR mainly include the influence of different WR structures on the detection signal, the quantitative analysis of LF type defects, the formulation of testing standards, and the relationship between LMA, LF defects and the strength of WRs (details can be found in reference [78]). Only by solving the quantitative detection of LF type defects and the relationship between defects and the strength of WRs, can the service life of WRs be truly and accurately predicted by non-destructive testing.
(2) In the field of WR optical detection, the existing research has a small amount of data, namely no big data sets, and the algorithm is with shallow structure. It has some problems such as limited

excavation ability, low computational efficiency and poor robustness. It is difficult to carry out high-speed detection, high efficiency and reliable identification, therefore can not be directly applied to real-time state health monitoring of WR. By introducing the deep learning [79–82] method into the WRs damage detection/monitoring, it is expected to achieve real-time, reliable and accurate damage identification and location, thus realizing the detection and monitoring of WR surface damage [83].

(3) The ultrasonic guided wave method, acoustic emission method and other detection methods are still in the stage of theoretical research and laboratory, and there is still a certain distance away from practical application, due to the limitation of technical level and the method itself. UGW and AE method are difficult to be used in actual dynamic detection, and the signal-to-noise ratio needs to be further improved; Eddy current method is less studied and affected by skin effect. Its reliability needs further study; Ray Method's development prospects are relatively limited compared with other methods because of the expensive instruments and the potential impact of radiation on the human body. Only by further deepening theoretical and experimental research and solving more of the above difficult problems can we promote the progress of these detection methods.

5. Summaries and Prospects

To promote the development of non-destructive detection method for WRs, we present an overview of non-destructive damage detection methods for WRs in this paper. Some summaries are listed as follows:

(1) The types of WR damages and their causes were introduced in detail, including wire breakage, wear, rust, deformation, and fatigue. They are divided into two types: local flaw and loss of metallic cross-sectional area.

(2) The development status of several important detection methods including electromagnetic detection, optical detection, ultrasonic guided wave method, acoustic emission detection, eddy current detection, and ray detection was reviewed, and their advantages and disadvantages were compared and summarized. On the whole, electromagnetic detection method has gradually been applied in practice. Optical method has shown great potential for application, while other methods are still in the laboratory stage.

In addition, some research trends and potential future research directions are given as follows:

(1) Electromagnetic detection method: the influence of WR structure on electromagnetic signal is expected to be studied; to solve the quantitative detection of LF type defect, the relationship between LF type defect and electromagnetic signal characteristics should be further investigated; the relationship between defect and WR strength should be further determined through a large number of experimental studies, so as to achieve real and accurate prediction of WR service life; and relevant standards should be further improved and developed combining the research results with the practical experience.

(2) Optical detection method: methods to reduce or eliminate the influence of oil and light on the surface of WRs should be further designed; deep learning method is expected to be introduced because of its strong data mining ability, and more efficient, accurate, and robust algorithm than the traditional machine learning should be designed; to make the designed algorithm more robust, a big data set of WR surface defects with different WR types and different sizes (diameter, wire number) is expected to be established to cover all sample space.

(3) Other detection methods: theoretical and experimental research, especially ultrasonic guided wave method, acoustic emission method, eddy current detection method and ray detection method, should be further deepened, and the intrinsic relationship between detection methods and the respective parameters of WR is expected to be explored; at the same time, relevant detection equipment should be designed and optimized to promote the development of related technologies.

(4) Comprehensive application: a promising avenue for research is to combine various methods to characterize and detect damages from multiple dimensions. For example, the combination of electromagnetic method and optical method can not only grasp the size of effective cross-sectional area of WRs, the types of internal and external damages in real time, but also intuitively master its surface morphology characteristics, so as to provide more valuable parameters for the WR health evaluation.

Although there are many issues and challenges, as new technologies/methods and algorithms are introduced, it is believed that this field will have a more and more prospective future and one day it will be possible to truly realize efficient and accurate non-destructive detection of WRs.

We believe that this review has synthesized individual pieces of information on non-destructive damage detection of steel WRs and has provided a comprehensive reference for these readers who are interested in this research field.

Author Contributions: Conceptualization, P.Z. and G.Z.; Formal analysis, P.Z., G.Z., Z.Z., and Z.H.; Funding acquisition, G.Z. and P.Z.; Investigation, P.Z., Z.Z., X.D., C.T., and Z.H.; Supervision, G.Z., Z.Z., C.T., and Z.H.; Writing—original draft, P.Z.

Funding: This work was supported by the National Key Research and Development Program of China (grant number 2016YFC0600905), by the Postgraduate Research & Practice Innovation Program of Jiangsu Province (grant number KYCX19_2140) and Postgraduate Research & Practice Innovation Program of China University of Mining and Technology (grant number KYCX19_2140), by the Jiangsu Provincial Outstanding Youth Fund of China (grant number BK20180033), by the National Natural Science Foundation of China (grant number 51575513), and by the Project Funded of the Priority Academic Program Development of Jiangsu Higher Education Institutions (PAPD).

Acknowledgments: Thanks for the journal editors for their kind invitation to present this feature article. We wish to thank the authors of all references for their outstanding contributions in this field. At the same time, the authors would like to thank all reviewers and editors for their constructive comments.

Conflicts of Interest: The authors declare no conflict of interest.

References

1. Mouradi, H.; Barkany, A.E.; Biyaali, A.E. Steel wire ropes failure analysis: Experimental study. *Eng. Fail. Anal.* **2018**, *91*, 234–242. [CrossRef]
2. Henao, H.; Fatemi, S.M.J.R.; Capolino, G.A.; Sieg-Zieba, S. Wire rope fault detection in a hoisting winch system by motor torque and current signature analysis. *IEEE Trans. Ind. Electron.* **2011**, *58*, 1727–1736. [CrossRef]
3. Peng, Y.X.; Chang, X.D.; Sun, S.S.; Zhu, Z.C.; Gong, X.S.; Zou, S.Y.; Xu, W.X.; Mi, Z.T. The friction and wear properties of steel wire rope sliding against itself under impact load. *Wear* **2018**, *400*, 194–206. [CrossRef]
4. Krešák, J.; Peterka, P.; Kropuch, S.; Novákb, L. Measurement of tight in steel ropes by a mean of thermovision. *Measurement* **2014**, *50*, 93–98. [CrossRef]
5. Yuan, F.; Hu, B.L.; Zhou, Z.J. An analysis on the research status quo and prospects of defect detection methods of wire ropes. *Mach. Des. Manuf.* **2010**, *2*, 260–262. [CrossRef]
6. Čereška, A.; Zavadskas, E.K.; Bucinskas, V.; Podvezko, V.; Sutinys, E. Analysis of steel wire rope diagnostic data applying multi-criteria methods. *Appl. Sci.* **2018**, *8*, 260. [CrossRef]
7. Zhang, J.W.; Tan, X.J.; Zheng, P.B. Non-destructive detection of wire rope discontinuities from residual magnetic field images using the hilbert-huang transform and compressed sensing. *Sensors* **2017**, *17*, 608. [CrossRef] [PubMed]
8. Tan, J.W. *Principle and Technology of Steel Wire Rope Safety Detection*; Science Press: Beijing, China, 2009.
9. Yang, S.Z.; Kang, Y.H.; Chen, H.G.; Yuan, J.M. *Electromagnetic Nondestructive Testing of Wire Ropes*; Machinery Industry Press: Beijing, China, 2017.
10. Gu, W.; Chu, J.X. A transducer made up of fluxgate sensors for testing wire rope defects. *IEEE Trans. Instrum. Meas.* **2002**, *51*, 120–124. [CrossRef]
11. Tan, X.J. Research on Non-Destructive Testing of Ferromagnetic Components with Weak Magnetic Imaging: In a Case of Wire Rope. Master's Thesis, Henan University of Science and Technology, Henan, China, 2018.

12. Norouzi, E.; Ravanbod, H. Optimization of the flux distribution in magnetic flux leakage testing. *Mater. Eval.* **2010**, *3*, 360–364. [CrossRef]
13. Zhao, M.; Zhang, D.L. Magnetic flux leakage of typical defect of wire rope based on FE simulation. *Nondestr. Test.* **2009**, *31*, 177–180.
14. Lenard, S.; Atherton, D.L. Calculation of the effects of anisotropy on magnetic flux leakage detector signals. *IEEE Trans. Magn.* **1996**, *32*, 1905–1909. [CrossRef]
15. Krzywosz, K. Comparion of electromagnetic techniques for nondestructive inspetion of ferromagnetic tubing. *Mater. Eval.* **1990**, *48*, 42–45.
16. Kalwa, E.; Piekarski, K. Design of inductive sensors for magnetic testing of steel ropes. *NDT E Int.* **1991**, *24*, 328. [CrossRef]
17. Dutta, S.M.; Ghorbel, F.H.; Stanley, R.K. Simulation and analysis of 3-D magnetic flux leakage. *IEEE Trans. Magn.* **2009**, *45*, 1966–1972. [CrossRef]
18. Trevino, D.A.G.; Dutta, S.M.; Ghorbel, F.H.; Karkoub, M. An improved dipole model of 3-D magnetic flux leakage. *IEEE Trans. Magn.* **2016**, *52*, 1–7. [CrossRef]
19. Sun, Y.H.; Kang, Y.H. Magnetic mechanisms of magnetic flux leakage nondestructive testing. *Appl. Phys. Lett.* **2013**, *103*, 184104. [CrossRef]
20. Gao, G.H.; Lian, M.J.; Xu, Y.G.; Qin, Y.N.; Gao, L. The effect of variable tensile stress on the MFL signal response of defective wire ropes. *INSIGHT* **2016**, *58*, 135–141. [CrossRef]
21. Coramik, M.; Ege, Y. Discontinuity inspection in pipelines: A comparison review. *Measurement* **2017**, *111*, 359–373. [CrossRef]
22. Kang, Y.H.; Li, Z.J.; Yang, Y.; Qiu, C. Mini-micro sensor and device for wire rope MFL testing. *Nondestr. Test.* **2014**, *36*, 11–15.
23. Wang, H.Y.; Hua, G.; Tian, J. Research on detection device for broken wires of coal mine-hoist cable. *J. Chin. Univ. Mini. Technol.* **2007**, *17*, 376–381. [CrossRef]
24. Jomdecha, C.; Prateepasen, A. Design of modified electromagnetic main-flux for steel wire rope inspection. *NDT E Int.* **2009**, *42*, 77–83. [CrossRef]
25. Wang, H.Y.; Xu, Z.; Hua, G.; Tian, J.; Zhou, B.B.; Lu, Y.H.; Chen, F.J. Key technique of a detection sensor for coal mine wire ropes. *Min. Sci. Technol.* **2009**, *19*, 170–175. [CrossRef]
26. Song, K.; Chen, C.; Kang, Y.H.; Li, J.J.; Ren, J.L. Mechanism study of AC-MFL method using U-shape inducer. *Chin. J. Sci. Instrum.* **2012**, *33*, 1980–1985. [CrossRef]
27. Yan, X.L.; Zhang, D.L.; Pan, S.M.; Zhao, E.C.; Gao, W. Online nondestructive testing for fine steel wire rope in electromagnetic interference environment. *NDT E Int.* **2017**, *92*. [CrossRef]
28. Kaur, A.; Gupta, A.; Aggarwal, H.; Arora, K.; Garg, N.; Sharma, M.; Sharma, S.; Aggarwal, N.; Sapra, G.; Goswamy, J.K. Non-destructive evaluation and development of a new wire rope tester using parallely magnetized NdFeB magnet segments. *J. Nondestr. Eval.* **2018**, *37*, 61. [CrossRef]
29. Sun, Y.H.; Wu, J.B.; Feng, B.; Kang, Y.H. An opening electric-mfl detector for the ndt of in-service mine hoist wire. *IEEE Sens. J.* **2014**, *14*, 2042–2047. [CrossRef]
30. Liu, X.C.; Xiao, J.W.; Wu, B.; He, C.F. A novel sensor to measure the biased pulse magnetic response in steel stay cable for the detection of surface and internal flaws. *Sens. Actuators A Phys.* **2018**, *269*, 218–226. [CrossRef]
31. Yan, X.L.; Zhang, D.L.; Zhao, F. Improve the signal to noise ratio and installation convenience of the inductive coil for wire rope nondestructive testing. *NDT E Int.* **2017**, *92*, 221–227. [CrossRef]
32. Fedorko, G.; Molnár, V.; Ferková, Ž.; Peterka, P.; Krešák, J.; Tomašková, M. Possibilities of failure analysis for steel cord conveyor belts using knowledge obtained from non-destructive testing of steel ropes. *Eng. Fail. Anal.* **2016**, *67*, 33–45. [CrossRef]
33. Sun, Y.H.; Liu, S.W.; Li, R.; Ye, Z.J.; Kang, Y.H.; Chen, S.B. A new magnetic flux leakage sensor based on open magnetizing method and its on-line automated structural health monitoring methodology. *Struct. Health Monit.* **2015**, *14*, 583–603. [CrossRef]
34. Xu, F.Y.; Wang, X.S.; Wu, H.T. Inspection method of cable-stayed bridge using magnetic flux leakage detection: Principle, sensor design, and signal processing. *J. Mech. Sci. Technol.* **2012**, *26*, 661–669. [CrossRef]
35. Wang, H.T.; Tian, J.; Meng, G.Y. A sensor model for defect detection in mine hoisting wire ropes based on magnetic focusing. *INSIGHT* **2017**, *59*, 143–148. [CrossRef]

36. Kim, J.W.; Park, S. Magnetic flux leakage sensing and artificial neural network pattern recognition-based automated damage detection and quantification for wire rope non-destructive evaluation. *Sensors* **2018**, *18*, 109. [CrossRef]
37. Tian, J.; Wang, H.Y. Research on magnetic excitation model of magnetic flux leakage for coal mine hoisting wire rope. *Adv. Mech. Eng.* **2015**, *7*, 1–11. [CrossRef]
38. Zhou, J.Y.; Tian, J.; Wang, H.Y.; Li, Y.M.; Wu, M. Numerical simulation of magnetic excitation based on a permanent magnet co-directions array sensor. *INSIGHT* **2018**, *60*, 568–574. [CrossRef]
39. Singh, W.S.; Rao, B.P.C.; Mukhopadhyay, C.K.; Jayakumar, T. GMR-based magnetic flux leakage technique for condition monitoring of steel track rope. *INSIGHT* **2011**, *53*, 377–381. [CrossRef]
40. Liu, X.C.; Wang, Y.J.; Wu, B.; Gao, Z.; He, C.F. Design of tunnel magnetoresistive-based circular MFL sensor array for the detection of flaws in steel wire rope. *J. Sens.* **2016**, *2016*. [CrossRef]
41. Lu, T.; Zhang, L.H. *Wire Rope Inspector*; China University of Mining and Technology Press: Xuzhou, China, 2007; pp. 74–78.
42. Zhou, P.; Zhou, G.B.; He, Z.Z.; Tang, C.Q.; Zhu, Z.C.; Li, W. A novel texture-based damage detection method for wire ropes. *Measurement* **2019**. under review.
43. Vallan, A.; Molinari, F. A vision-based technique for lay length measurement of metallic wire ropes. *IEEE Trans. Instrum. Meas.* **2009**, *58*, 1756–1762. [CrossRef]
44. Yaman, O.; Karakose, M. Auto-correlation based elevator rope monitoring and fault detection approach with image processing. In Proceedings of the 2017 International Artificial Intelligence and Data Processing Symposium (IDAP), Malatya, Turkey, 16–17 September 2017.
45. Sun, H.X.; Zhang, Y.H.; Luo, F.L. Texture segmentation and boundary recognition of wire rope images in complicated background. *Acta Photonica Sinica* **2010**, *39*, 1666–1671. [CrossRef]
46. Platzer, E.S.; Nägele, J.; Wehking, K.H.; Denzler, J. HMM-based defect localization in wire ropes–a new approach to unusual subsequence recognition. Pattern Recognition. In Proceedings of the Dagm Symposium, Jena, Germany, 9–11 September 2009; pp. 442–451.
47. Platzer, E.S.; Wehking, K.H.; Denzler, J. On the suitability of different features for anomaly detection in wire ropes. In Proceedings of the International Conference on Computer Vision, Imaging and Computer Graphics, Lisboa, Portugal, 5–8 February 2009; pp. 296–308.
48. Wacker, E.S.; Denzler, J. Enhanced anomaly detection in wire ropes by combining structure and appearance. *Pattern Recognit. Lett.* **2013**, *34*, 942–953. [CrossRef]
49. Wacker, E.S.; Denzler, J. An analysis-by-synthesis approach to rope condition monitoring. In Proceedings of the International Symposium on Visual Computing, Las Vegas, NV, USA, 29 November–1 December 2010; pp. 459–468.
50. Sanchez-Brea, L.M.; Siegmann, P.; Rebollo, M.A.; Bernabeu, E. Optical technique for the automatic detection and measurement of surface defects on thin metallic wires. *Appl. Opt.* **2000**, *39*, 539–545. [CrossRef] [PubMed]
51. Laguerre, L.; Treyssede, F. Non-destructive evaluation of seven-wire strands using ultrasonic guided waves. *Eur. J. Environ. Civil Eng.* **2011**, *15*, 487–500. [CrossRef]
52. Raisutis, R.; Kazys, R.; Mazeika, L.; Zukauskas, E.; Samaitis, V.; Jankauskas, A. Ultrasonic guided wave-based testing technique for inspection of multi-wire rope structures. *NDT E Int.* **2014**, *62*, 40–49. [CrossRef]
53. Liu, Z.H.; Zhao, J.C.; Wu, B.; Zhang, Y.N.; He, C.F. Configuration optimization of magnetostrictive transducers for longitudinal guided waves inspection in seven-wire steel strands. *NDT E Int.* **2010**, *43*, 484–492. [CrossRef]
54. Treyssède, F.; Laguerre, L. Investigation of elastic modes propagating in multi-wire helical waveguides. *J. Sound Vib.* **2010**, *329*, 1702–1716. [CrossRef]
55. Vanniamparambil, P.A.; Khan, F.; Hazeli, K.; Cuadra, J.; Schwartz, E.; Kontsos, A.; Bartoli, I. Novel optico-acoustic nondestructive testing for wire break detection in cables. *Struct. Control Health Monit.* **2013**, 1319–1350. [CrossRef]
56. Xu, J.; Wu, X.J.; Sun, P.F. Detecting broken-wire flaws at multiple locations in the same wire of prestressing strands using guided waves. *Ultrasonics* **2013**, *53*, 150–156. [CrossRef]
57. Raisutis, R.; Kazys, R.; Mazeika, L.; Samaitis, V.; Zukauskas, E. Propagation of ultrasonic guided waves in composite multi-wire ropes. *Materials* **2016**, *9*, 451. [CrossRef]
58. Tse, P.W.; Rostami, J. Advanced signal processing methods applied to guided waves for wire rope defect detection. In *AIP Conference Proceedings*; AIP Publishing: College Park, MD, USA, 2016; Volume 1706, p. 030006.
59. Li, H.; Wang, W.T.; Zhou, W.S. Fatigue damage monitoring and evolution for basalt fiber reinforced polymer materials. *Smart Struct. Syst.* **2014**, *14*, 307–325. [CrossRef]

60. Hoon, S.; Gyuhae, P.; Jeannette, R.W.; Nathan, P.L.; Charles, R.F. Wavelet-based active sensing for delamination detection in composite structures. *Smart Mater. Struct.* **2003**, *13*, 153. [CrossRef]
61. Drummond, G.; Watson, J.F.; Acarnley, P.P. Acoustic emission from wire ropes during proof load and fatigue testing. *NDT E Int.* **2007**, *40*, 94–101. [CrossRef]
62. Bai, W.; Chai, M.; Li, L.; Li, Y.; Duan, Q. Acoustic emission from elevator wire ropes during tensile testing. In *Advances in Acoustic Emission Technology: Proceedings of the World Conference on Acoustic Emission-2013*; Springer: New York, NY, USA, 2013; pp. 217–224.
63. Casey, N.F.; Taylor, J.L. Acoustic emission of steel wire ropes. *Wire Ind.* **1984**, *51*, 79–82.
64. Casey, N.F.; Taylor, J.L. Evaluation of wire ropes by AE techniques. *Brit. J. Non. Destruct. Test.* **1985**, *27*, 351–356.
65. Casey, N.F.; Wedlake, D.; Taylor, J.L.; Holford, K.M. Acoustic detection of wire rope failure. *Wire Ind.* **1985**, *52*, 307–309.
66. Casey, N.F.; Taylor, J.L.; Holford, K.M. Wire break detection during tensile fatigue testing of 40 mm wire rope. *Brit. J. Non. Destruct. Test.* **1985**, *30*, 338–341.
67. Casey, N.F.; White, H.; Taylor, J.L. Frequency analysis of the signals generated by the failure of constituent wires of a wire rope. *NDT E Int.* **1989**, *56*, 583–586. [CrossRef]
68. Casey, N.F.; Laura, P.A.A. A review of the acoustic-emission monitoring of wire rope. *Ocean Eng.* **1997**, *24*, 935–947. [CrossRef]
69. Shao, Y.B.; Yu, D.; Wang, S.; Zhu, Z.M.; Yin, W.Q. Quantitative method to detect wire break in wire rope by acoustic emission techniques. *J. Northeast. Univ. Nat. Sci.* **1999**, *20*, 130–132.
70. Ding, Y.; Reuben, R.L.; Steel, J.A. A new method for waveform analysis for estimating AE wave arrival times using wavelet decomposition. *NDT E Int.* **2004**, *37*, 279–290. [CrossRef]
71. Li, S.; Wu, Y.; Shi, H. A novel acoustic emission monitoring method of cross-section precise localization of defects and wire breaking of parallel wire bundle. *Struct. Control Health Monit.* **2019**, *26*, e2334. [CrossRef]
72. Li, H.; Huang, Y.; Chen, W.L.; Ma, M.L.; Tao, D.W.; Ou, J.P. Estimation and warning of fatigue damage of FRP stay cables based on acoustic emission techniques and fractal theory. *Comput. Aided Civil Infrastruct. Eng.* **2011**, *26*, 500–512. [CrossRef]
73. Cao, Q.S.; Liu, D.; Zhou, J.H.; Zhou, J.M. Non-destructive and quantitative detection method for broken wire rope. *Chin. J. Sci. Instrum.* **2011**, *32*, 787–794. [CrossRef]
74. Cao, Q.S.; Liu, D.; He, Y.H.; Zhou, J.H.; Codrington, J. Nondestructive and quantitative evaluation of wire rope based on radial basis function neural network using eddy current inspection. *NDT E Int.* **2012**, *46*, 7–13. [CrossRef]
75. Zhang, Y.C.; Xu, G.Y.; Cui, J. System used to detect steel wire rope based on X-ray digital imaging technology. *Image Technol.* **2008**, *2*, 33–39.
76. Peng, P.C.; Wang, C.Y. Use of gamma rays in the inspection of steel wire ropes in suspension bridges. *NDT E Int.* **2015**, *75*, 80–86. [CrossRef]
77. Cao, Y.N.; Zhang, D.L.; Xu, D.G. The state-of-art of quantitative nondestructive testing of wire ropes. *Nondestruct. Test.* **2005**, *27*, 91–95.
78. Wu, P.; Hua, H.Y. Discussion on the existing problems of steel wire rope nondestructive testing. *Nondestruct. Test.* **2017**, *39*, 65–68. [CrossRef]
79. Zhou, P.; Zhou, G.B.; Zhu, Z.C.; Tang, C.Q.; He, Z.Z.; Li, W.; Jiang, F. Health monitoring for balancing tail ropes of a hoisting system using a convolutional neural network. *Appl. Sci.* **2018**, *8*, 1346. [CrossRef]
80. Guo, X.J.; Chen, L.; Shen, C.Q. Hierarchical adaptive deep convolution neural network and its application to bearing fault diagnosis. *Measurement* **2016**, *93*, 490–502. [CrossRef]
81. Zhao, R.; Yan, R.Q.; Chen, Z.H.; Mao, K.Z.; Wang, P.; Gao, X. Deep learning and its applications to machine health monitoring. *Mech. Syst. Signal Process.* **2019**, *115*, 213–237. [CrossRef]
82. Guo, X.J.; Shen, C.Q.; Chen, L. Deep fault recognizer: An integrated model to denoise and extract features for fault diagnosis in rotating machinery. *Appl. Sci.* **2017**, *7*, 41. [CrossRef]
83. Zhou, P.; Zhou, G.B.; He, Z.Z.; Zhu, Z.C.; Tang, C.Q.; Li, W. A novel deep learning-based damage monitoring and diagnosis method for wire ropes. *Mech. Syst. Signal Process.* **2019**. under review.

© 2019 by the authors. Licensee MDPI, Basel, Switzerland. This article is an open access article distributed under the terms and conditions of the Creative Commons Attribution (CC BY) license (http://creativecommons.org/licenses/by/4.0/).

Article

Lock-In Thermography and Ultrasonic Testing of Impacted Basalt Fibers Reinforced Thermoplastic Matrix Composites

Simone Boccardi [1,2], Natalino Daniele Boffa [1], Giovanni Maria Carlomagno [1], Giuseppe Del Core [2], Carosena Meola [1,*], Ernesto Monaco [1], Pietro Russo [3] and Giorgio Simeoli [3]

1. Department of Industrial Engineering—Aerospace Division, University of Naples Federico II, Via Claudio, 21, 80125 Napoli, Italy
2. Department of Science and Technology, University of Naples Parthenope, Centro Direzionale, Isola C4, 80143 Napoli, Italy
3. Institute for Polymers, Composites and Biomaterials, National Council of Research, 80078 (NA) Pozzuoli, Italy
* Correspondence: carmeola@unina.it

Received: 27 June 2019; Accepted: 25 July 2019; Published: 26 July 2019

Featured Application: The term "new material" may raise enthusiasm as well as skepticism at the same time. The acceptance of new material is generally linked to its characteristics with respect to the commonly used materials such as better performance, the solution to an open question, non- or less-polluting, cheaper, easy to handle, etc. Of course, these characteristics must be certified; therefore, the production should be accompanied by effective testing techniques and procedures. The scope of this work is to contribute to assessing the effectiveness of non-destructive testing through the examination of both lock-in thermography and ultrasonic testing techniques applied to basalt-based composites.

Abstract: Basalt fibers are receiving increasing consideration because they seem to be adequate as reinforcement of composites and to comply with the environmental safeguard rules. However, many factors affect the performance of composite material, demanding specific testing; one may be performance assessment under impact tests. The attention of the present work is focused on the detection of impact damage in basalt-based composites with two non-destructive testing techniques: lock-in thermography (LT) and ultrasonic testing (UT). Two different types of materials are considered which both include basalt fibers as reinforcement but two different matrices: Polyamide and polypropylene. Polypropylene is used either pure or modified with the addition of a coupling agent; the latter improves the fiber/matrix interface strength, giving in practice, a material of different characteristics. Specimens are first subjected to low-velocity impact tests and then non-destructively examined with the two above mentioned techniques. The obtained results are analyzed and compared to highlight the advantages and limitations of the two techniques to detect impact damage in basalt-based composites. Both techniques seem effective for the inspection of polyamide/basalt composite; in particular, there is a general agreement between results. Conversely, UT seems not suitable for the inspection of polypropylene/basalt composites because of their superficial porosity, while lock-in thermography is effective also for this type of composite material.

Keywords: basalt fibers; polyamide; polypropylene; composites; impact damage; lock-in thermography; ultrasonic testing

1. Introduction

The development of new materials is an increasingly topical issue of great interest to both the academic and the industrial communities. This is because of many reasons. Amongst them, the performance of a product depends mostly on the material which it is made of, and a demand for materials of superior characteristics. Another important question is the environmental safeguard, which requires the development of more environmentally friendly materials. Therefore, the attention is ever more shifting from petrochemical resources to more natural (e.g., vegetables) substances. The hope is to be able to get eco-friendly composite materials which include both the matrix and reinforcement of natural derivation and which perform better, or at least similar to the most common composites of petrochemical derivation. This seems not completely achievable yet; what is instead possible is to reduce the problems of waste disposal. In this context, thermoplastic matrix-based composites (TC), thanks to their potential recyclability after their life-cycle, offer some advantages over their thermoset counterparts [1]. Another step forward may be to use natural fibers as reinforcement of thermoplastic matrices. A convenient reinforcement may be basalt.

Basalt is available in nature in volcanic rocks and can be reduced in fibers, which are well suited to be used as reinforcement of both thermoset and thermoplastic matrices to create different types of composite materials [2].

Based on the investigation till now carried out [3,4], the obtained composites seem to have good features, which make them comparable, or superior, to the most commonly used composites. The basalt-based composites could be adequate for the construction of some aircraft parts and of unmanned aerial vehicles like drones, which are ever more applied in different sectors such as surveillance and remote inspection. However, such adequacy should be ascertained since many factors affect the performance of a composite material. This one must be subjected to many tests prior to its introduction in the market for the construction of any goods. Amongst them, it is generally important to assess the material resistance to an impact load. The latter can be inadvertently induced during either manufacturing, through the impact of falling objects, or in service, and can cause barely visible damage. Therefore, the availability of effective non-destructive testing (NDT) techniques is fundamental to get information on the damage induced especially by low energy impact.

Today, many different NDT techniques are available, but none can be considered as superior and very effective; in fact, every technique has its inherent limitations. A good practice is to choose the most adequate technique to the specific application but, frequently, an effective routine may be to use more than one technique in the light of a data integration/fusion approach.

The attention of the present work is focused on the detection of low energy impact damage in basalt-based composites with two techniques: lock-in thermography (LT) and ultrasonic testing (UT). Both techniques were already used successfully to detect the impact damage of carbon-based composites [5] and now are considered again for the inspection of basalt-based composites. Basalt fibers are used as reinforcement of a polyamide matrix and of a polypropylene (PP) one, which is used either pure, or modified with the addition of a coupling agent; the latter improves the fiber/matrix interface strength, giving in practice, a material of different characteristics. Therefore, three different materials are inspected. Specimens are first subjected to low velocity/energy impact tests and then non-destructively examined with the two above mentioned techniques. The obtained results are compared to highlight the advantages and limitations of the two techniques within the inspection of basalt-based composites. In addition to the Introduction, this work is organized into several sections. Section 2, titled Materials and Methods, includes a description of the used materials and specimens and how they are impacted as well as of the two non-destructive techniques: lock-in thermography and ultrasonic testing. Section 3 reports Results separately in Section 3.1 for polyamide matrix-based specimen and in Section 3.2 for polypropylene matrix-based specimens. Section 4 titled: Data Discussion and Concluding Remarks will close the paper.

2. Materials and Methods

2.1. Description of Specimens

A woven basalt fibers fabric, plain weave type, with a specific mass of 210 g/m^2, from Incotelogy, GmbH is used as reinforcement of two types of matrix: Polypropylene and polyamide. The first is polypropylene, Hyosung Topilene PP J640 (MFI@230 °C, 2.16 kg: 10 g/10 min; Songhan Plastic Technology Co., Ltd., Shanghai, China), which is used pure, or modified by adding 2% in weight of Polybond 3000 (PP-g-MA, MFI@190 °C, 2.16 kg: 405 g/10 min; 1.2% in weight of maleic anhydride, by Chemtura, Philadelphia, PA, USA). The other is polyamide (PA6) Lanxess Durethan B30S-000000 (MFI@260 °C, 5 kg: 102 g/10 min). Then, three types of specimens, named: PPB, PC2B, and PA6B are prepared with a pure PP (PPB), a modified one (PC2B), and a polyamide (PA6B) matrix. For the last, the neat interface without any coupling agent has been chosen even if there is mention in the literature of the enhancement introduced by silane coupling [6].

Each specimen includes 18 balanced basalt fabric layers 0°/90° symmetrically arranged with respect to the middle plane of the laminate ([(0/90)$_9$]$_s$ configuration), with a basalt fiber content of 50% by volume (the actual relative percentages of fiber and matrix evaluated according to ASTM D 3171-04, Test Method II). The percentage of 50% has been chosen as an optimal value through preliminary tests. Laminates are obtained by alternating layers of matrix films (PP, PC2, PA6) with basalt fibers fabric by the hand lay-up film-stacking technique and compacted with the aid of a compression molding machine (model P400E, Collin GmbH, Germany) under pre-optimized molding conditions (Figure 1). Each specimen is 300 mm × 300 mm with a target thickness of 3 mm.

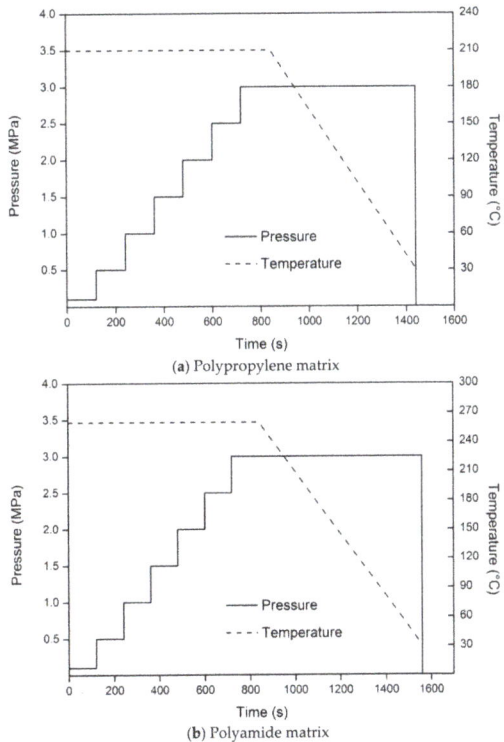

Figure 1. Molding compression pressure and temperature against time for polypropylene (PP) based (**a**) and polyamide (PA6B) based (**b**) matrices.

Some specific details of investigated specimens in terms of code, thickness, and composition are summarized in Table 1.

Table 1. Some specimens details.

Code	Matrix	Thickness (mm)
PA6B	Polyamide	3.0
PPB	Pure polypropylene	3.0
PC2B	Polypropylene added with 2% maleic anhydride	3.0

2.2. Impact Tests

Impact tests are carried out with a modified Charpy pendulum with a hemispherical shaped hammer nose, 12.7 mm in diameter. Each specimen is placed inside a special fixture which includes two large plates, each having a window 12.5 cm × 7.5 cm to allow for the contact with the hammer from one side and likely optical view by an infrared imaging device from the other one [7]. The impact energy varied between 5 and 15 J and is set by suitably adjusting the falling height of the Charpy arm. The choice of the impact energy was to induce mostly barely visible damage to different levels without perforation.

2.3. Non-Destructive Evaluation

After impact two non-destructive testing techniques are used: lock-in thermography and ultrasonic testing.

2.3.1. Lock-In Thermography

The test setup includes the impacted specimen, the infrared camera, and two halogen lamps (1 kW each) for thermal stimulation of the specimen [8]. The used infrared camera is the SC6000 (Flir systems), which is equipped with a QWIP detector, working in the 8–9 µm infrared band, NEDT <35 mK, spatial resolution 640 × 512 pixels full-frame, pixel size 25 µm × 25 µm and with a windowing option linked to frequency frame rate and temperature range. The camera is equipped with the lock-in module that drives the halogen lamps to generate a sinusoidal thermal wave of selectable frequency f and with the IRLock-In© software for data analysis.

Lock-in thermography is a well-known technology and will not be herein described in detail; only a sketch of the test setup (Figure 2) and some basics are recalled for easy reading. The thermal wave, delivered to the specimen surface, propagates inside the material and gets reflected when it reaches parts where the heat propagation parameters change (in-homogeneities). The reflected wave interacts with the surface wave producing an oscillating interference pattern, which can be measured in terms of either temperature amplitude, or phase angle φ, and represented as an amplitude, or phase, image, respectively. The basic link of the thermal diffusion length μ to the heating frequency f and to the mean material thermal diffusivity coefficient α is via the relationship:

$$\mu = \sqrt{\frac{\alpha}{\pi f}}. \tag{1}$$

The depth range for the amplitude image is given by μ, while the maximum depth p, which can be reached for the phase image, corresponds to 1.8 μ [9–12]. In general, it is preferable to reduce data in terms of phase image because of its insensitivity to either non-uniform heating or local variations of emissivity over the monitored surface. The material thickness, which can be inspected, depends on the wave period (the longer the period, the deeper the penetration) and on the material thermal diffusivity.

As stated in Equation (1), to check the material conditions at a given depth, both the knowledge of the thermal diffusivity and the correct choice of the heating frequency are fundamental parameters. The thermal diffusivity can be evaluated with either the lock-in technique itself or with flash thermography.

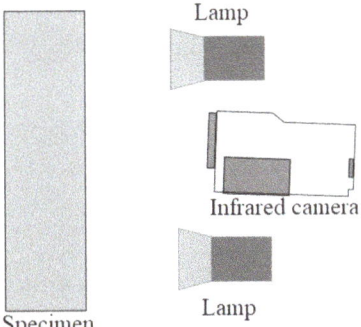

Figure 2. Scheme of the lock-in thermography setup.

2.3.2. Ultrasonic Testing

Ultrasonic non-destructive inspection is performed with an ultrasonic flaw detector Olympus OmniScan SX with phased array unit (Figure 3). The Omniscan detector is equipped with a phased array 16:64PR probe and with an ultrasonic (UT) conventional channel allowing for pulse-echo (PE), pitch-catch, or time-of-flight diffraction (TOFD) methods of inspection [13]. To perform the inspection the phased array probe elements are pulsed simultaneously, or with a time lag, in order to promote ultrasonic beamforming that is the constructive interference of multiple-beam components into a single wavefront traveling in the desired direction. Correspondingly, in the acquiring mode, the receiver is able to combine the signals (echoes) coming from multiple elements (reflectors) into a single representation. Exploiting the phasing technology that allows beam shaping and steering, it is possible to generate a wide variety of ultrasonic beam profiles with a single probe assembly, and, at the same time, to dynamically set the beam steering to perform electronic scans.

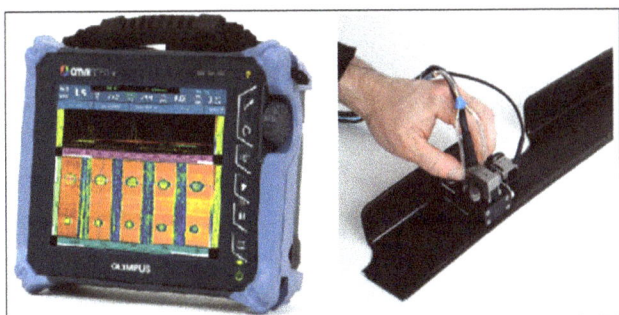

Figure 3. Some details of the ultrasonic (UT) testing setup.

Phased array ultrasonic instruments, whose operating principle relies on the physics that governs sound wave propagation, apply high-frequency sound waves to either detect buried anomalies or measure the thickness of a testing article [14]. The capability to produce multiple transducer paths within one probe represents a great advantage in defects detection and visualization [15]. Phased array imaging allows seeing relative point by point changes and multidirectional defect responses, which help in flaw detection, discrimination, and sizing [16].

In particular, the ultrasonic device records two parameters of the reflected echo: The amplitude and the echo time of flight with respect to a zero point (pulse transit time). The echo time of flight, in turn, is correlated with the depth or distance of the reflector, exploiting the sound velocity knowledge of the tested material and the simple relationship:

$$\text{Distance} = \text{velocity} \times \text{time}. \qquad (2)$$

The fundamental presentation of ultrasonic waveform data is the A-scan graph, in which amplitude and transit time of the echo are plotted on a grid where the vertical axis represents the amplitude and the horizontal axis represents the time of flight.

A different plotting way is in the form of B-scan, which visualizes the depth of reflectors (a defect/damage) with respect to their position along the scanning axis. The inspected test article thickness is plotted as a function of time or position while the probe is moved along the upper part surface to provide material depth profile. The correlation of echoes data with the transducer positions allows a through-thickness representation of the material and to link track data with the specific inspected areas.

The probe position tracking is performed by the use of an electromechanical encoder, a small wheel connected to the probe, that enables the position and orientation of the probe along the scanning axis to be recorded with the echoes amplitude data (A-scan) allowing the 2D reconstructions. A useful 2D presentation option is the C-scan graph, in which the data are displayed in a top or planar view of the test piece, something similar to an x-ray image, where color represents the signal amplitude or the reflector depth at each point of the inspected test piece.

In the present work, tests are performed by means of an encoded 5 MHz, phased array transducer with 64 active elements arranged in a linear array with a pitch of 1 mm and a straight wedge. The system calibration is performed by considering the specimen thickness as a reference, so no calibration blocks are used.

3. Results

Results are presented as phase images for LT and as C-scan and B-scan images, amplitude top and sectional views respectively, for ultrasonic testing (UT); data are compared to highlight the advantages and limitations of the two techniques. As the first and most important observation, the comparison between LT and UT is done only for the PA6B specimen. This is because, from preliminary tests, the surface of both PPB and PC2B specimens are found to be permeable to the ultrasonic coupling medium (water, or gel); this has made the C-scan inspection impossible to perform in consequence of the background noise amplification due to the liquid ingress effect, which blinds the ultrasonic device. Therefore, these two specimens are inspected only with LT.

3.1. Polyamide Matrix-Based Specimen

Firstly, a photo of the PA6 specimen taken after impact is reported in Figure 4. As indicated, the specimen was subjected to four impacts with energies: 5, 9, and 15 J and with two impacts at the same $E = 9$ J. The location of each impact is also recognizable from the light local surface discoloration.

Three phase images taken from the impacted side (shown in Figure 4) are reported in Figure 5; a reference target of 19 mm × 8 mm, clearly visible on each image, allows for fast estimation of the damage size. From these phase images it is possible to see that:

- There are dark stains in correspondence of the four impacts.
- The size of dark stains increases with the impact energy.
- Going in-depth, the stains first expand by decreasing f from 0.53 Hz (Figure 5a) to 0.36 Hz (Figure 5b), and after contracting to their initial size to a further reduction of f to 0.19 Hz (Figure 5c).

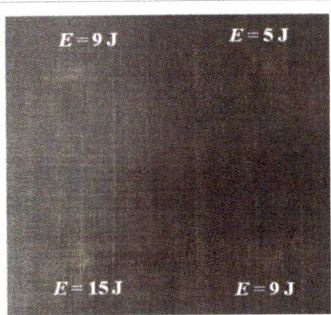

Figure 4. Photo of the polypropylene added with 2% maleic anhydride (PA6B) specimen with indicated energy and location of the impacts.

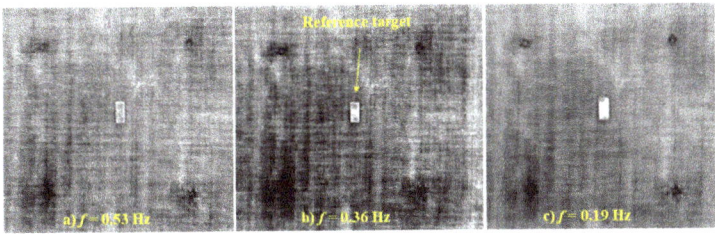

Figure 5. Phase images taken on the impacted side of the PA6B specimen at three heating frequencies; (a) $f = 0.53$ Hz; (b) $f = 0.36$ Hz; (c) $f = 0.19$ Hz (see Figure 2 for impact energies location).

In particular, the dark zone centered at the impact point of $E = 15$ J (Figure 5a) assumes for $f = 0.36$ Hz (Figure 5b) a well H shaped configuration being representative of the delamination evolution along the vertical fiber's direction. Such a vertical displacement may be explicable owing to the fact that, during the impact, the specimen was positioned inside the lodge with the vertical direction of the fibers (as shown in Figure 5b) along the longer side of the lodge's window (12.5 cm × 7.5 cm). In addition, it is possible to see threadlike structures resembling cracks in the matrix which are most pronounced around the impact point of 9 J on the bottom (Figure 5b). However, apart from the clearly visible impact coupled stains, there are other dark zones mostly present on the phase image taken at $f = 0.36$ Hz (Figure 5b), which are distributed away from the impact points and which may indicate that material in-homogeneities likely occurred during the manufacturing process. All of these dark structures fade going more in-depth; they remain clearly evident for $f = 0.19$ Hz (Figure 5c) mostly around the impact points.

What is observed in Figure 5b is practically confirmed by the C-scan image of Figure 6a. In fact, the black color of Figure 5b is replaced by the dark blue color of Figure 6a to highlight the indentation of each impact and the most important damage.

In particular, it is confirmed the presence of delamination along the direction of the vertical fibers and it is also confirmed the presence of material in-homogeneity around impacts at 15 J, 9 J (on the right bottom), and 5 J and far away from the impact points. In addition, the dark structures around the impact at 5 J of Figure 5 are better depicted as blue-green structures in Figure 6. The comparison between Figures 5 and 6 helps to classify such structures as material in-homogeneities rather than impact-induced damage. As already evidenced, such in-homogeneities display their best contrast for $f = 0.36$ Hz (Figure 5b) meaning that they are mostly located in the second half of the specimen thickness towards the bottom. It is worth noting that around the impact points in-homogeneities are present for lower f (Figure 5c) and so extend to the entire thickness. This set of evidence, probably exacerbated by impact effects, are confirmed in the ultrasonic B-scan sections (Figure 6b), where it is possible to see wide porosity areas through the stratification close to the opposite surface of the impacted side.

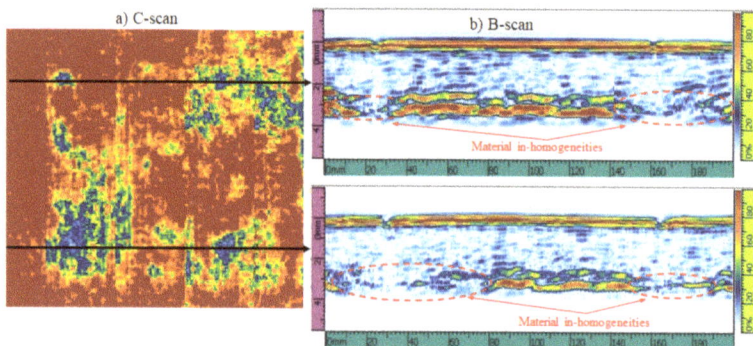

Figure 6. C-scan (**a**) and B-scan (**b**) images taken on the impacted side of the PA6B specimen (see Figure 4 for impact energies location).

3.2. Polypropylene Matrix-Based Specimens

The Figures 7 and 8 show phase images, taken from the impacted side, of the two specimens PPB (Figure 7) and PC2B (Figure 8), which were subjected to three impacts at 5 J, 9 J, and 15 J as indicated on the first image (a) of both Figures 8 and 9. Figure 9 shows a comparison between phase images taken at $f = 0.19$ Hz from the rear side (opposite to impact) of the three specimens PA6B (Figure 9a), PPB (Figure 9b), and PC2B (Figure 9c).

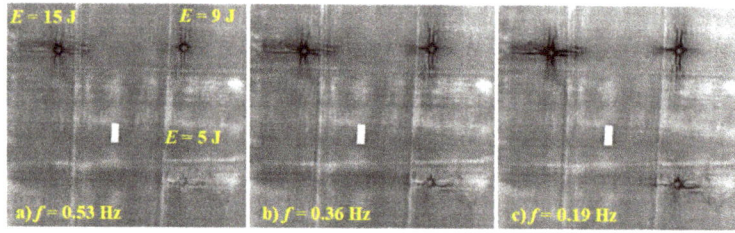

Figure 7. Phase images taken on the impacted side of the pure polypropylene (PPB) specimen at three heating frequencies; (**a**) $f = 0.53$ Hz; (**b**) $f = 0.36$ Hz; (**c**) $f = 0.19$ Hz.

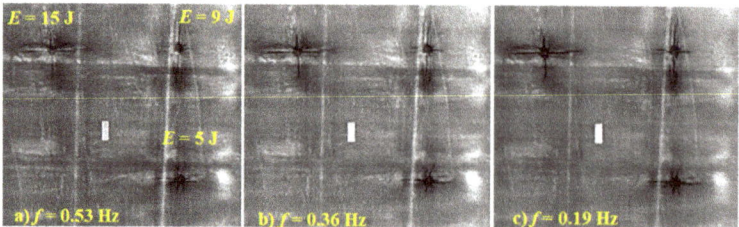

Figure 8. Phase images taken on the impacted side of the PC2B specimen at three heating frequencies; (**a**) $f = 0.53$ Hz; (**b**) $f = 0.36$ Hz; (**c**) $f = 0.19$ Hz.

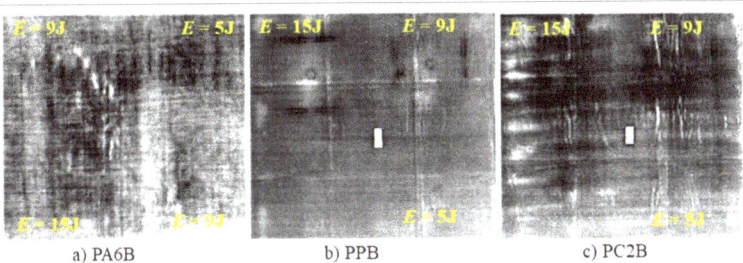

Figure 9. Phase images taken at $f = 0.19$ Hz on the rear side of the three specimens; (**a**) PA6B; (**b**) PPB; (**c**) PC2B.

As a general observation, the occurred damage appears in the form of a cross-shaped structure with a central ring. The cross's branches indicate the main deformation directions while the ring may be assumed to coincide with the imprint of the impactor. The cross is more accentuated for the highest impact energy of 15 J and achieves its best contrast at $f = 0.19$ Hz (Figure 7c Figure 8c). As a main difference between the two specimens, the cross is depicted by fan-like branches in the PPB specimen (Figure 7) and by almost straight lines in the PC2B one (Figure 8). Such a difference is a symptom of different behavior of the two materials under the impact. It is worth noting that the superimposed layers of the PPB specimen are not fully rigidly connected at their interface but are free to somehow mutually slide. Instead, the stronger bond entailed by the grafting agent in the PC2B specimen prevents slipping effects of layers and allows for bending and deformation through the entire thickness. This may entail wrinkling in the layers furthest from the impact point as shown in Figure 9c.

3.3. Measurement of Damage Extension

The extension of the occurred impact damage is evaluated by considering its elongation along horizontal (D_H) and vertical (D_V) directions as represented in Figure 10. These values are extracted from the phase images through a correspondence pixels/mm by considering the reference target.

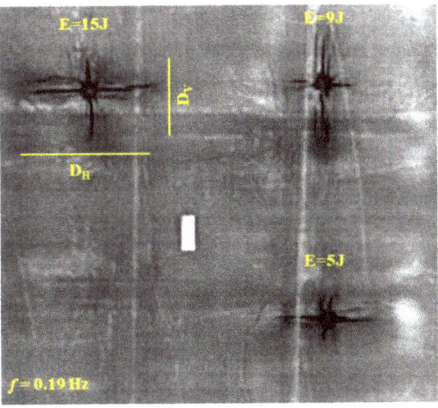

Figure 10. Indication of D_H and D_V over a phase image.

D_H and D_V values are supplied only for PPB (Table 2) and PC2B (Table 3) specimens for which the most important damage develops along well-defined directions.

Table 2. Values of D_H and D_V for the PPB specimen through LT testing.

E (J)	D_H (mm)	D_V (mm)
5	40.0	24.0
9	46.0	54.0
15	75.4	42.0

Table 3. Values of D_H and D_V for the PC2B specimen through LT testing.

E (J)	D_H (mm)	D_V (mm)
5	58.0	24.0
9	38.0	74.6
15	75.4	44.0

Instead, as already shown and described, the PA6B specimen displays patchy damage which cannot be simplified with its displacement in the two directions only; therefore, no measurements are reported for this specimen. Coming back to Tables 2 and 3 it is possible to see the tendency towards longer cuts on the PC2B specimen. This is reliable because improving the interface bond makes the material more brittle and prone to fractures.

4. Data Discussion and Concluding Remarks

The obtained results lend themselves to several considerations:

- The investigated materials are susceptible to defects formation during manufacturing. These defects may include non-uniform distribution of matrix and may be ascribed to non-uniform distribution of temperature and application of pressure during the compression cycle. This is likely to occur because of the compression performed in a press.
- Both lock-in thermography and ultrasonic testing can discover either impact damage, or manufacturing defects in PA6 matrix-based specimens with a general data agreement. This is well documented by a comparison between Figures 3 and 4.
- Polypropylene/basalt specimens being hydrophilic get soaked with the coupling gel and cannot be inspected with gel-based UT. Instead, lock-in thermography, acting in a remote way without any contact, is well suited and effective to detect both manufacturing defects and impact damage, also in polypropylene-based specimens.
- Specimens involving PA6 as a matrix display better mechanical properties and react to impact with less extensive damage with respect to specimens involving polypropylene as a matrix. This because both polyamide and basalt fibers have polar chemical structure, which enables proper interface adhesion even in the absence of a coupling agent. Instead, the a-polar nature of polypropylene requires the addition of coupling agents and/or any prior treatment of fibers to assure good interfacial adhesion.
- The presence of the coupling (compatibilizing) agent has no significant effects on the extension of the impact damage with regards to the branches length on the superficial layer but mostly affects the deformation way of the underlying layers. More specifically, a stronger interface adhesion entails a crumple effect with stretching of the bottom layer. This effect is evident looking at the phase images taken from the rear of specimens PA6B (Figure 9a) and PC2B (Figure 9c) in comparison with the phase image of the specimen PPB (Figure 9b), which appear almost flat.

On the whole, and by taking into account the points listed above, it seems preferable to build composites by embedding basalt fibers inside a polyamide matrix since the obtained composites seem to possess better characteristics and can be inspected with both LT and UT techniques. However, in some specific applications, polypropylene-based composites may be preferable relying on the use of infrared thermography for non-destructive evaluation. It is also worth noting that by means of specific measures it is possible to improve manufacturing and guarantee surface waterproofing allowing testing with liquid-coupled probes. Of course, the choice of materials involves many other factors such as: performance, suitability, green aptitude, easy inspection, costs, etc., leading to a compromise.

Author Contributions: Conceptualization: All authors; specimens preparation: G.S.; impact tests: N.D.B. and E.M.; investigation with LT: S.B., and C.M.; software and data reduction: S.B.; investigation with UT: N.D.B., E.M.; data analysis and discussion: C.M., S.B., N.D.B., and G.S.; writing—original draft preparation: C.M.; revision: G.M.C.; all authors revised and approved the final version.

Funding: This research received no external funding.

Conflicts of Interest: The authors declare no conflict of interest.

References

1. Collier, M.C.; Baird, D.G. Separation of a thermotropic liquid crystalline polymer from polypropylene composites. *Polym. Compos.* **1999**, *20*, 423–435. [CrossRef]
2. Fiore, V.; Scalici, T.; Di Bella, G.; Valenza, A. A review on basalt fibre and its composites. *Compos. Part B Eng.* **2015**, *74*, 74–94. [CrossRef]
3. Lopresto, V.; Leone, C.; De Iorio, I. Mechanical characterisation of basalt fibre reinforced plastic. *Compos. Part B Eng.* **2011**, *42*, 717–723. [CrossRef]
4. Sarasini, F.; Tirillò, J.; Valente, M.; Valente, T.; Cioffi, S.; Iannace, L. Sorrentino, Effect of basalt fiber hybridization on the impact behavior under low impact velocity of glass/basalt woven fabric/epoxy resin composites. *Compos. Part A Appl. Sci. Manuf.* **2013**, *47*, 109–123. [CrossRef]
5. Meola, C.; Boccardi, S.; Carlomagno, G.M.; Boffa, N.D.; Monaco, E.; Ricci, F. Nondestructive evaluation of carbon fibre reinforced composites with infrared thermography and ultrasonics. *Compos. Struct.* **2015**, *134*, 845–853. [CrossRef]
6. Vikas, G.; Sudheer, M. A Review on Properties of Basalt Fiber Reinforced Polymer Composites. *Am. J. Mater. Sci.* **2017**, *7*, 156–165.
7. Boccardi, S.; Boffa, N.D.; Carlomagno, G.M.; Del Core, G.; Meola, C.; Russo, P.; Simeoli, G. Inline monitoring of basalt-based composites under impact tests. *Compos. Struct.* **2019**, *210*, 152–158. [CrossRef]
8. Meola, C.; Boccardi, S.; Carlomagno, G.M. *Infrared Thermography in the Evaluation of Aerospace Composite Materials*; Woodhead Publishing Print Book: Sawston, UK, 2016; 180p, ISBN 9781782421719.
9. Busse, G. Optoacoustic phase angle measurement for probing a metal. *Appl. Phys. Lett.* **1979**, *35*, 759–760. [CrossRef]
10. Letho, A.; Jaarinen, J.; Tiusanen, T.; Jokinen, M.; Luukkala, M. Magnitude and phase in thermal wave imaging. *Electron. Lett.* **1981**, *17*, 364–365.
11. Beaudoin, J.-L.; Merienne, E.; Danjoux, R.; Egee, M. Numerical system for infrared scanners and application to the subsurface control of materials by photothermal radiometry. *Proc. SPIE* **1985**, *590*, 287–292.
12. Busse, G.; Wu, D.; Karpen, W. Thermal wave imaging with phase sensitive modulated thermography. *J. Appl. Phys.* **1992**, *71*, 3962–3965. [CrossRef]
13. *Advances in Phased Array Ultrasonic Technology Applications*; Manual; Olympus Scientific Solutions Americas (OSSA): Waltham, MA, USA, 2007.
14. Ensminger, D.; Bond, L.J. *Ultrasonics: Fundamentals, Technologies, and Applications*, 3rd ed.; CRC Press: Boca Raton, FL, USA, 2011.
15. Taheri, H.; Hassen, A.A. Nondestructive Ultrasonic Inspection of Composite Materials: A Comparative Advantage of Phased Array Ultrasonic. *Appl. Sci.* **2019**, *9*, 1628. [CrossRef]
16. Drinkwater, B.W.; Wilcox, P.D. Ultrasonic arrays for non-destructive evaluation: A review. *NDT&E Int.* **2006**, *39*, 525–541.

© 2019 by the authors. Licensee MDPI, Basel, Switzerland. This article is an open access article distributed under the terms and conditions of the Creative Commons Attribution (CC BY) license (http://creativecommons.org/licenses/by/4.0/).

Communication

Toward Creating a Portable Impedance-Based Nondestructive Testing Method for Debonding Damage Detection of Composite Structures

Wongi S. Na * and Ki-Tae Park

Sustainable Infrastructure Research Center, Korea Institute of Civil Engineering & Building Technology, Gyeonggi-Do 10223, Korea
* Correspondence: wongi@kict.re.kr; Tel.: +82-31-910-0155

Received: 29 May 2019; Accepted: 29 July 2019; Published: 5 August 2019

Abstract: Debonding detection of composite structures is a vital task as such damage weakens the structure leading to a failure. As adhesive bonding is a more preferable choice over the conventional mechanical fastening method, detecting debonding as early as possible could minimize the overall maintenance costs. For this reason, a vast amount of research in the nondestructive testing field is being conducted as we speak. However, most of the methods may require well-trained experts or heavy equipment. In this study, the piezoelectric (PZT) material-based method known as the electromechanical impedance technique is investigated to seek the possibility of making the technique very cheap and portable by temporarily attaching the sensor. Furthermore, ANSYS simulation studies using smaller PZT patches as small as 0.1 mm × 0.1 mm are simulated to investigate the impedance signatures acquired from the simulations. The results show the possibility of using smaller PZT patches compared to the conventional PZT sizes of 10 mm × 10 mm for a successful damage identification process.

Keywords: debonding; composite damage; electromechanical impedance; piezoelectric; nondestructive testing; FEM simulation

1. Introduction

Detecting damage in composite materials is becoming more and more important with the increase in applications for composites. Up to date, there are many nondestructive testing (NDT) methods for evaluating the structural integrity of a target structure [1–9]. For composites, since bonding of two composite parts with adhesives is the preferred choice over the conventional mechanical fastening approach, detecting debonding is a crucial factor when maintaining a composite structure. Some of the recent studies in this area include effects of sensor debonding failure on mathematical representation of smart composite laminate [10], piezoelectric wafer guided wave-based debonding of carbon fiber reinforced polymer (CFRP) overlay in CFRP-reinforced concrete structures [11], and laser ultrasonic guided waves for detecting debonding of multilayered bonded composites [12]. Although various NDT methods are available for detecting damage, most of the methods require well-trained experts or heavy equipment to perform the procedure. For this reason, a low-cost, portable NDT system that is easy to use can be an important item in the field of NDT. Moreover, with advance in technology, efforts have been made by combining NDT with IoT (Internet of Things) technology, and a well-summarized work in this area can be found in [13].

The electromechanical impedance (EMI) technique is one of the NDT techniques that use a single piezoelectric (PZT) transducer to act as both actuator and sensor. The technique involves measuring the impedance below 500 kHz and monitoring the changes in the signature for damage identification. The one-dimensional equation first introduced by Liang et al. [14] shows that the electrical impedance of the PZT transducer is directly related to the mechanical impedance of the host structure as shown below. Here, the electrical admittance, $Y(\omega)$, is a combined function of the mechanical impedance of the host structure, $Z_s(\omega)$, and the PZT transducer, $Z_a(\omega)$, respectively. This proves that any change in the host structure can be monitored by measuring $Y(\omega)$. Other variables in the equation, I, V, ω, a, ε_{33}^T, δ, d_{3x}, $\overline{Y}_{xx}^E$, are the PZT output current, PZT input voltage, input frequency, geometric constant, dielectric constant, loss tangent, piezoelectric constant, and Young's modulus, respectively.

$$Y(\omega) = i\omega a \left(\varepsilon_{33}^T (1 - i\delta) - \frac{Z_s(\omega)}{Z_s(\omega) + Z_a(\omega)} d_{3x}^2 \overline{Y}_{xx}^E \right) \qquad (1)$$

Currently, there are three ways of conducting the EMI technique. The most-used one is by using an impedance analyzer (e.g., Agilent 4924a), which can measure impedance over 1 MHz depending on specification. However, such equipment can be very costly and weigh over 10 kg making it impractical for field use. Another way of conducting the EMI technique is by combining a function generator with an oscilloscope and using a simple circuit proposed by Peairs et al. [15]. Using this approach, one can conduct the EMI technique for a cost less than 25% compared to the conventional approach, and various authors have tested the reliability of this low-cost version. The cheapest way of conducting the EMI technique is by using the AD5933 evaluation board commercialized by Analog Devices Co. The small size of the device gives the possibility of creating a portable system for conducting the EMI technique. The cost of the AD5933 evaluation board can be as low as under 2% of the cost of the conventional impedance analyzer, which makes the EMI technique a very cheap technique to perform. However, one of the downsides of the device is that it can only measure impedance up to 100 kHz. Since one needs a signature with multiple peaks (e.g., resonance) to successfully identify damage, up to 100 kHz of frequency range might be not enough and could result in an impedance signature without any peak, resulting in failing to detect any damage. Especially for composite materials, due to its non-homogenous property, an impedance signature without any resonance can be commonly seen. To overcome this problem, the concept of sandwiching a metal disc between the PZT transducer and the host structure was proposed in [16], where its reliability was tested using glass fiber epoxy laminates.

To create a portable impedance measuring system, the way of attaching the PZT transducer must be changed. Since the EMI technique requires the PZT patch to be permanently attached to the host structure, one must alter this so that the PZT transducer can be temporarily attached and detached for multiple EMI measurements. The easiest way to achieve this would be to use a double-sided tape with the PZT–metal EMI transducer; the reliability and performance of this idea are tested in this study. In addition to the test, simulation studies on using PZT in smaller sizes are introduced.

2. Evaluation of the PZT–Metal Transducer Using Double-Sided Tape

The idea of the PZT–metal transducer proposed in [16] used a commercialized epoxy adhesive for attachment to the host structure. Thus, to make the PZT–metal transducer re-attachable, a simple idea of using a double-sided tape (purchased from www.alpha.co.kr) is evaluated in this section against debonding and crack damage. Figure 1 shows the experimental setup to evaluate the performance of the PZT–metal transducer with double-sided tape attachment. The AD5933 evaluation board is connected to the laptop as shown in the figure where the positive and negative wires from the board are attached to the PZT–metal transducer by soldering. More details on the PZT–metal transducer with its damage detection performance can be found in [16]. For evaluating the performance of the PZT–metal transducer, two glass fiber epoxy laminates of 200 mm × 70 mm with 0.4 mm thickness were adhered to each other using a commercialized epoxy (Loctite Quick-set). After fully curing the epoxy for 48 h at room temperature, the impedance signature was measured from 25 to 75 kHz in 100 Hz steps to be used as a reference signature. Then, debonding was achieved by inserting a chisel tip between the two composite plates and seperating them apart 10 mm at a time until 100 mm debonding was achieved. For each 10 mm of debonding, impedance sigantures were measured. After 100 mm debonding was achieved, the final impedance signature was measured with complete debonding of the bottom composite plate. The reason for this was to observe the difference in the impedance signature subjected to debonding that occurs away from the PZT transducer and which that occurs right below the PZT transducer. To check the reliability of the double-sided tape attachment approach, five identical tests (Test 1, Test 2, Test 3, Test 4, and Test 5) were conducted.

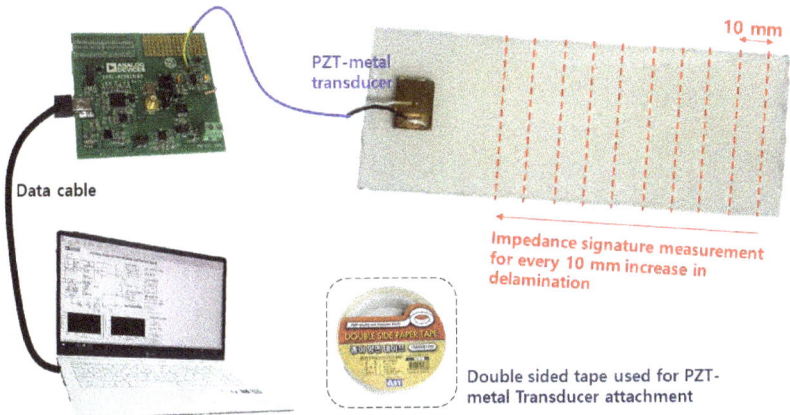

Figure 1. Experimental setup for the piezoelectric (PZT)–metal transducer experiment.

Observing Figure 2a, which represents the impedance signatures from Test 1, debonding has caused the amplitude peak (at 60 kHz) to decrease with increase in debonding area. However the amplitude remains virtually the same for the frequency range at 35 kHz where change is very difficult to see. However, when bottom composite plate is removed, the impedance signature peak increases dramatically at the 60 kHz resonance range. This experimentally proves that signature change is significant when debonding occurs below the PZT–metal transducer. Figure 2b shows the limitation of using the double-sided tape as the five impedance signatures seen in this figure were acquired by removing and reattaching the PZT–metal transducer onto the same area of the composite plate. The change is not significant but such change in signatures can cause false alarms when no damage has been experienced by the structure.

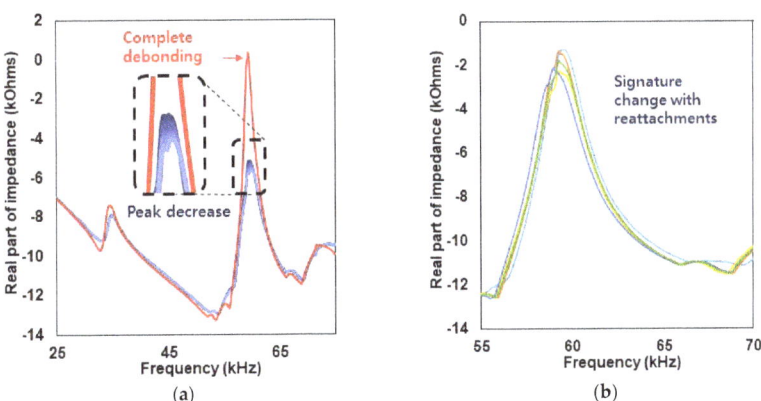

Figure 2. Impedance signature results for: (**a**) Debonding experiment; (**b**) re-attachable performance of the PZT–metal transducer.

For evaluating the damage detection performance, the variations between the reference and corresponding signatures (after damage) were quantified using root-mean-square deviation (RMSD) in equations below. In the equations, $Re(Z_i)$ and $Re(Z_i^o)$ represents the real part of the reference and the corresponding impedance signatures, respectively.

$$\text{RMSD} = \sqrt{\sum_N [Re(Z_i) - Re(Z_i^o)]^2 / \sum_N [Re(Z_i^o)]^2} \qquad (2)$$

Looking at Figure 3, the RMSD values with the reference signature being the intact case, the RMSD value was 0.6% with 1 cm debonding and increased to 1.37% with 5 cm debonding. However the RMSD values decreased down to 0.97% with 9 cm debonding. This shows the limitation of debonding detection using the PZT–metal transducer, where it seems that the highest value that can be obtained from this test is 1.37%. With full removal of the bottom composite plate, the RMSD resulted in 9.06%. Table 1 summarizes all the five test results including the data from Figure 3. Although the same tests were conducted, the RMSD results were different. With Test 2, all the RMSD values were below 1% and it was difficult to observe the increasing trend subjected to increase in debonding length. However, debonding below the PZT–metal transducer resulted in 10.29% once again proving that this NDT technique is effective at finding damage below the sensor. For Test 3, the increasing trend was clearly seen with increase in debonding length as the RMSD value started from 0.66% and rose to 2.31% with 8 cm debonding length. Here, the RMSD value with full removal of the bottom plate resulted in 10.98%, which shows that the reliability of finding defects located under a PZT–metal transducer is very effective. Tests 4 and 5 show similar results compared to Test 2 where all the RMSD values were below 1%, where the values generally increase with increase in debonding length. With full removal of the plate, the RMSD values were 11.9% and 15.32% for Tests 4 and 5, respectively.

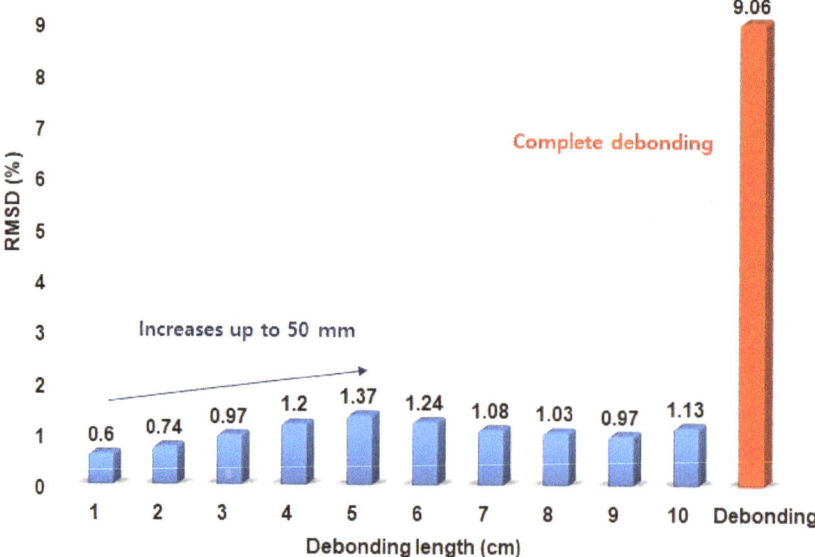

Figure 3. Root-mean-square deviation (RMSD) values for the debonding experiment and total removal of the bottom plate for Test 1.

Table 1. Summarized RMSD (%) values for the five tests.

	Test 1	Test 2	Test 3	Test 4	Test 5
10 mm	0.6	0.65	0.66	0.12	0.17
20 mm	0.74	0.68	1.03	0.22	0.34
30 mm	0.97	0.61	1.25	0.28	0.5
40 mm	1.2	0.56	1.55	0.38	0.49
50 mm	1.37	0.58	1.77	0.53	0.28
60 mm	1.24	0.57	2.05	0.66	0.34
70 mm	1.08	0.48	2.14	0.68	0.46
80 mm	1.03	0.44	2.31	0.72	0.51
90 mm	0.97	0.65	2.3	0.76	0.55
100 mm	1.13	0.81	2.3	0.8	0.89
Complete debonding	9.06	10.27	10.98	11.9	15.32

3. FEM (Finite Element Method) Simulation for Smaller PZT Patches

From the previous section, although the PZT–metal transducer was able to detect debonding damage of composite plates, the impedance signature changed dramatically when debonding occurred below the attached PZT transducer. For this reason, various tests with smaller PZT sizes should be tested to validate the suitability of using smaller PZT transducers, as using smaller PZT could detect small debonding areas. However, using PZT in smaller sizes (compared to the conventional size of 10 mm × 10 mm with 0.5 mm thickness) would be very difficult to perform experimentally. For an example, it would be impossible to perform an EMI technique using a 0.1 mm square PZT as soldering of the positive and negative sides of the PZT would be difficult. Thus, to first check the reliability of the simulation, commercial FEM software ANSYS Workbench was used with coupled field analysis. Here, finite element couples the effects of interrelated physics within the element matrices making the electromechanical impedance simulation possible. In addition, the convergence criteria in ANSYS was unaltered as the program itself produced acceptable performance compared to the experimental result. Using ANSYS, a model representing the PZT–metal transducer used in the previous section with 6407 nodes and 1096 elements was created as shown in Figure 4. The average element quality was 0.959. In addition, the properties used for the PZT material can be seen in Table 2. The properties for the metal part of the PZT–metal transducer were selected from the ANSYS Workbench engineering database (stainless steel).

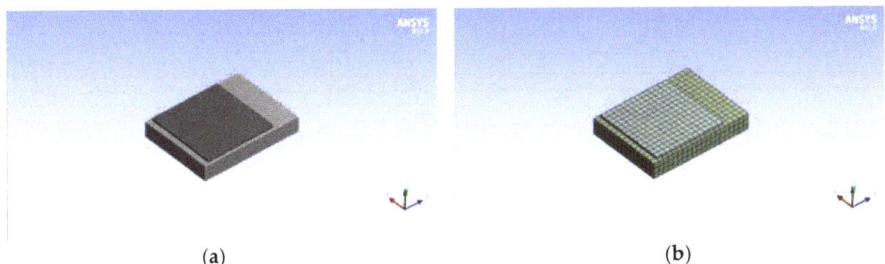

(a)   (b)

Figure 4. ANSYS FEM model: (a) PZT–metal; (b) mesh with 1096 elements.

Table 2. PZT properties for ANSYS simulation study.

		PSI-5A4E
Density		7800
Damping Ratio		0.0125
Stiffness Matrix [c^E]	$C_{11} = C_{22}$	152
	C_{12}	102
	$C_{13} = C_{23}$	100
	C_{33}	127
	$C_{44} = C_{55}$	21
	C_{66}	25
Piezoelectric Stress Matrix [e]	$e_{31} = e_{32}$	−5.5
	e_{33}	16.4
	$e_{24} = e_{15}$	12.4
Electric Permittivity Matrix [ε^s]	$\varepsilon_{11} = \varepsilon_{22}$	950
	ε_{33}	890

Figure 5a shows the actual impedance signature of the PZT–metal transducer in air using the AD5933 evaluation board, and Figure 5b shows the impedance signature from the ANSYS simulation. Comparing both results, although the shapes of the signature are slightly different, the resonance frequency range from the FEM simulation result matches very well with the actual experimental result. Two large resonance peaks exist, where the larger one is located at around 65 kHz for both results (Figure 5a,b), and the smaller resonance peak for both signatures is located near 40 kHz. Furthermore, the amplitude of the large peak is around twice the size of the smaller resonance peak.

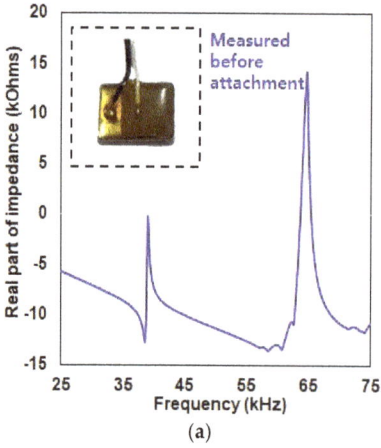
(a)

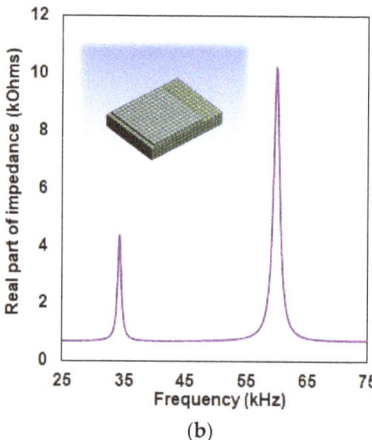
(b)

Figure 5. Impedance signature of the PZT–metal transducer before structure attachment: (a) By experiment; (b) by simulation.

Now that we had confirmed that the FEM simulation result was reliable, four more FEM models were created to evaluate the impedance signatures. PZT patches sizing from as small as 0.1–5 mm were created to investigate for resonance. Figure 6a shows the FEM model with a 0.1 mm × 0.1 mm PZT patch with 0.01 mm thickness attached to a square metallic material on the bottom (0.2 mm × 0.2 mm with 0.01 mm thickness), Figure 6b looks identical to Figure 6a but the size is 10 times larger. The PZT size is 1 mm × 1 mm with 0.1 mm thickness with the bottom metal plate of 2 mm × 2 mm with 0.1 mm thickness. Figure 6c has a 5 mm × 5 mm size PZT with 1 mm thickness and a metal size of 6 mm × 9 mm with 0.5 mm thickness. The last FEM model, Figure 6d, has the same dimension as the previous figure but with the thickness of the metal plate increased to 1.5 mm. All four models were meshed with element quality over 0.9 to achieve the best outcome before obtaining the simulation results.

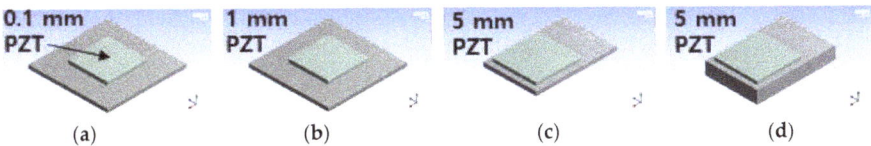

Figure 6. FEM model for smaller PZT sizes: (**a**) 0.1 mm PZT; (**b**) 1 mm PZT; (**c**) 5 mm PZT; (**d**) 5 mm PZT with thicker metal plate.

Figure 7 shows the impedance signatures acquired from the four FEM models in the frequency range 20–200 kHz. With the 0.1 mm square PZT patch (Figure 6a), there is no resonance at all. This suggests that using such a small PZT will possibly result in a failure when detecting damage unless resonance can be found over the 200 kHz frequency range. Observing the 1 mm square PZT patch (Figure 6b), it has the highest value compared to the rest of the signatures. Larger than the signatures with 5 mm square PZTs (Figure 6c,d). This shows that larger PZT size does not always result in bigger amplitude. The two impedance signatures for the last two FEM models have completely different frequency ranges. The thinner bottom metal layer FEM model resulted in multiple resonance range with five resonance peaks below 150 kHz, whereas the last FEM model with the thicker metal layer had two resonance peaks (very small one at 90 kHz and another one near 200 kHz). Although these simulation results were not verified with experimental results due to the fact that it is virtually impossible to conduct experiments with PZT sizes of 0.1 mm × 0.1 mm × 0.01 mm, the fact remains that PZT sizes smaller than 10 mm × 10 mm can be used to detect damage as long as resonance is found. Furthermore, changing the bottom metal plate can also create resonance to overcome the problem of resonance-free signatures, which may cause the EMI technique to fail at detecting damage.

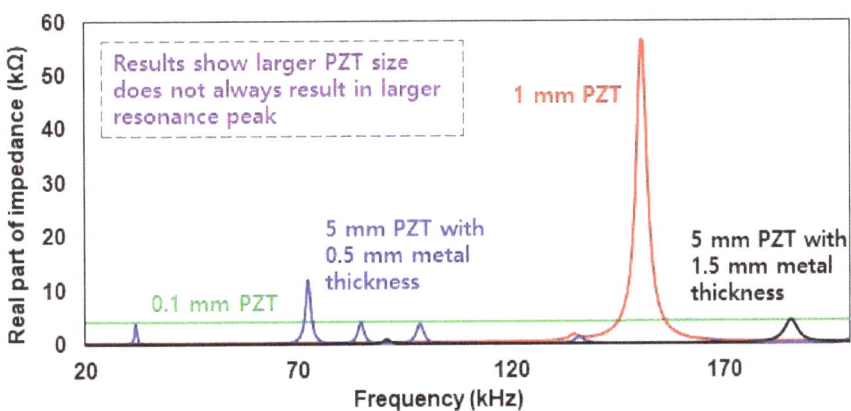

Figure 7. Simulation results for the impedance signatures of the four FEM models.

4. Conclusions

In this study, a work toward creating a cheap and portable nondestructive testing (NDT) method known as the electromechanical impedance (EMI) technique was carried out. Since the EMI technique requires the brittle PZT patch to be permanently attached to the structure for damage detection analysis, the idea of making the attachment temporarily was tested using a double-sided tape. Such an approach made the PZT–metal transducer very easy to be attached and detached. In addition, regardless of the temporary attachment approach, debonding damage of a glass fiber epoxy composite could be detected by monitoring the changes in the impedance signatures. However, this change was not as significant as the change subjected to debonding that occurred right below the PZT–metal transducer. This study experimentally showed that the debonding damage is best detected when it happens under the PZT patch. Therefore, to seek the possibility of detecting smaller debonding size, the size of the PZT patch had to be smaller than the conventional size of 10 mm × 10 mm with 0.5 mm thickness. Thus, the finite element analysis tool known as ANSYS Workbench was used to conduct simulations with smaller PZT patches as small as 0.1 mm × 0.1 mm with 0.01 mm thickness, virtually impossible to conduct experiment. Four different models were created with small PZT patches to find out that the size of the PZT was not the most important factor as the 1 mm square sized PZT patch showed resonance with highest peak amplitude.

Overall from this work, we found that the EMI technique can be made into a portable system, where the PZT transducer can be attached simply using a double-sided tape. Regardless of the damping effect, which may cause the impedance signatures to be less sensitive when subjected to damage, the results from this study have demonstrated its possibilities. Furthermore by conducting simulation studies, the PZT size can be further reduced for a successful debonding detection of composite structures.

Author Contributions: W.S.N. contributed in planning experiments and mainly writing the manuscript; K.-T.P. contributed in conducting experiments and analyzing data.

Funding: The research was supported by a grant from "Development of Safety Evaluation Techniques for Infrastructures by Micro–Macro Hybrid Monitoring Data (20190201-001)" funded by the Korea Research Institute of Standards and Science (KRISS).

Conflicts of Interest: The authors declare no conflict of interest.

References

1. Palumbo, D.; Tamborrino, R.; Galietti, U.; Aversa, P.; Tati, A.; Luprano, V.A.M. Ultrasonic analysis and lock-in thermography for debonding evaluation of composite adhesive joints. *NDT E Int.* **2016**, *78*, 1–9. [CrossRef]
2. Na, W.S. Low cost technique for detecting adhesive debonding damage of glass epoxy composite plate using an impedance based non-destructive testing method. *Compos. Struct.* **2018**, *189*, 99–106. [CrossRef]
3. Gholizadeh, S. A review of non-destructive testing methods of composite materials. *Procedia Struct. Integr.* **2016**, *1*, 50–57. [CrossRef]
4. Na, W.S. Distinguishing crack damage from debonding damage of glass fiber reinforced polymer plate using a piezoelectric transducer based nondestructive testing method. *Compos. Struct.* **2017**, *159*, 517–527. [CrossRef]
5. Babu, J.; Sunny, T.; Paul, N.A.; Mohan, K.P.; Philip, J.; Davim, J.P. Assessment of delamination in composite materials: A review. *Proc. Inst. Mech. Eng. Part B J. Eng. Manuf.* **2016**, *230*, 1990–2003. [CrossRef]
6. Meola, C.; Boccardi, S.; Carlomagno, G.M. A quantitative approach to retrieve delamination extension from thermal images recorded during impact tests. *NDT E Int.* **2018**, *100*, 142–152. [CrossRef]
7. Giri, P.; Mishra, S.; Clark, S.M.; Samali, B. Detection of gaps in concrete–metal composite structures based on the feature extraction method using piezoelectric transducers. *Sensors* **2019**, *19*, 1769. [CrossRef] [PubMed]
8. Qing, X.; Li, W.; Wang, Y.; Sun, H. Piezoelectric transducer-based structural health monitoring for aircraft applications. *Sensors* **2019**, *19*, 545. [CrossRef] [PubMed]
9. Xie, L.; Gao, B.; Tian, G.Y.; Tan, J.; Feng, B.; Yin, Y. Coupling pulse eddy current sensor for deeper defects NDT. *Sens. Actuators A Phys.* **2019**, *293*, 189–199. [CrossRef]
10. Khan, A.; Kim, H.S. Assessment of sensor debonding failure in system identification of smart composite laminates. *NDT E Int.* **2018**, *93*, 24–33. [CrossRef]
11. Li, J.; Lu, Y.; Guan, R.; Qu, W. Guided waves for debonding identification in CFRP-reinforced concrete beams. *Constr. Build. Mater.* **2017**, *131*, 388–399. [CrossRef]
12. Zhang, K.; Zhou, Z. Quantitative characterization of disbonds in multilayered bonded composites using laser ultrasonic guided waves. *NDT E Int.* **2018**, *97*, 42–50. [CrossRef]
13. Tokognon, C.A.; Gao, B.; Tian, G.Y.; Yan, Y. Structural health monitoring framework based on Internet of Things: A survey. *IEEE Internet Things J.* **2017**, *4*, 619–635. [CrossRef]

14. Liang, C.; Sun, F.P.; Rogers, C.A. Coupled electro-mechanical analysis of adaptive material systems—Determination of the actuator power consumption and system energy transfer. *J. Intell. Mater. Syst. Struct.* **1997**, *8*, 335–343. [CrossRef]
15. Peairs, D.M.; Park, G.; Inman, D.J. Improving accessibility of the impedance-based structural health monitoring method. *J. Intell. Mater. Syst. Struct.* **2004**, *15*, 129–139. [CrossRef]
16. Na, S.; Lee, H.K. Resonant frequency range utilized electro-mechanical impedance method for damage detection performance enhancement on composite structures. *Compos. Struct.* **2012**, *94*, 2383–2389. [CrossRef]

© 2019 by the authors. Licensee MDPI, Basel, Switzerland. This article is an open access article distributed under the terms and conditions of the Creative Commons Attribution (CC BY) license (http://creativecommons.org/licenses/by/4.0/).

Article

A Weighted Estimation Algorithm for Enhancing Pulsed Eddy Current Infrared Image in Ecpt Non-Destructive Testing

Hanchao Li, Yating Yu *, Linfeng Li and Bowen Liu

School of Mechanical and Electrical Engineering, University of Electronic Science and Technology of China, Chengdu 611731, China; lhc@uestc.edu.cn (H.L.); xhullf@163.com (L.L.); lbw9449@163.com (B.L.)
* Correspondence: wzwyyt@uestc.edu.cn; Tel.: +86-136-7813-9939

Received: 31 July 2019; Accepted: 13 September 2019; Published: 9 October 2019

Abstract: Non-destructive testing (NDT) plays a crucial role in large scale industrial production such as in the nuclear industry and bridge structures where even a small crack can lead to severe accidents. The pulsed eddy current infrared thermography testing method, as a classic non-destructive testing technology, is proposed to detect cracks in the presence of excitation sources that cause temperature changes in the vicinity of defects, which is higher than normal area. However, in the vicinity of the excitation sources, the temperature is higher than normal even if there is no defect. Traditional infrared image enhancing algorithms do not work efficiently when processing infrared images because the colors in the images represent the temperature. To address this, a novel algorithm is proposed in this paper. A weighted estimation algorithm is proposed because each pixel value has a strong relationship with its neighboring pixels. The value of each pixel is determined by calculating the values of its neighboring pixels with a specific step-size and the correlation coefficients between them. These coefficients are obtained by calculating the differences between the pixels. The experimental results indicated that the outline of the welding defect became significantly clearer after being processed using the proposed algorithm, which can eliminate the errors caused by the excitation source.

Keywords: non-destructive testing evaluation; infrared thermography testing; defect detection; image enhancement

1. Introduction

Defects detection such as in bridges and nuclear structures testing, especially the small and surface/subsurface cracks detection, is the main target of non-destructive testing (NDT) [1–3]. These kinds of defects such as welding defect, which can cause severe accidents in large scale industrial production and transportation, have been studied for many years [4]. The application of new materials for industrial processes make it difficult to detect defects using traditional NDT methods such as eddy current testing (ECT) technology. For instance, with the application of new materials in modern aircraft manufacturing, composites are widely used to manufacture the key components of the aircraft. As a consequence, infrared thermography testing which is an important branch of NDT [5–7], is gaining attention as an efficient method for defect detection in new materials [8–10]. Compared to other NDT technologies, infrared thermography testing has significant advantages. It can be used to detect defects for various materials and geometries [11]. It can also be used for the detection and location of wielding defects by collecting and processing infrared images [12–15].

Infrared thermography testing technology produces images that show the temperature distribution of the defects and the background [16].

The images are processed to collect as much information as possible to reconstruct the defects. Many researchers have done considerable amounts of work on the processing of infrared images [17,18].

The image processing approach can be classified into two categories namely spatial and frequency domain methods.

In spatial domain infrared image processing methods, the pixels of the image are manipulated directly. This is done by performing mathematical operations on each pixel and the surrounding pixels to obtain the gray value of the pixel. Common spatial domain image processing methods utilize sharp and smooth filters [19–21]. The fundamental purpose of the spatial domain processing method is to distinguish the defects from the background.

The frequency domain method transforms the image into frequency domain using a frequency transform methods such as the Fourier transform (FT) or wavelet transform (WT) [22]. The first step of the frequency domain image processing method is to convert the image into the frequency domain, after which the signal in the frequency domain is filtered. The frequency domain method yields low and high frequencies, which can be distinguished by a frequency transform of the infrared image. The low and the high frequency components are enhanced and suppressed respectively, such that the information of the defects in the image is made clearer.

However, most of the image processing methods only focus on the image itself, ignoring the significant differences between infrared and common images. In particular, infrared images do not show the actual scene being imaged. Instead, they show the distribution of temperature, which is influenced by excitation sources. In this paper, a method for enhancing infrared images is proposed. This method can eliminate the influence of the excitation source, and yield an infrared image with a clearer defect profile.

The essence of infrared thermography non-destructive testing technology is to obtain the temperature by measuring the amount of infrared radiation from the surface of an object. The common excitation sources of infrared thermography testing can be classified into three types: optical, electromagnetic, and mechanical excitation sources. In electromagnetic excitation sources, eddy current pulsed thermography (ECPT) is widely used as it can uniformly generate sufficient heat in a short time, and the heat it provides is generated from the eddy current induced from the excitation source. This allows the object to be heated evenly and makes the outline of the defect in the infrared image clearer [23].

The operating principle of eddy current pulsed thermography can be explained as follows. A high-frequency alternating current is applied to the excitation coil, which induces an eddy current on the surface and inside the conductive object. The inducted eddy current flows in the conductive object. The conduction of the eddy current is influenced by the size, shape, and position of the defect, as the eddy current is forced to flow around the defect [24].

Figure 1 shows the spreading process of the eddy current in an object containing a defect. The distribution of the eddy current is influenced by the defect. The eddy current is much denser the edge of the defect than it is in other regions without defects.

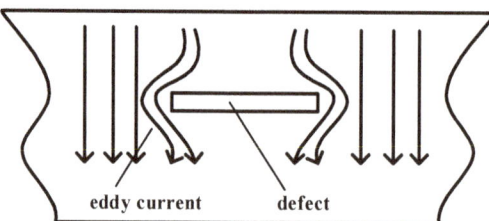

Figure 1. Eddy current at the edge of the defect.

The distribution of heat generated according to Joule's Law is uniform, because the distribution of eddy current is uniform. Joule's law is as follows:

$$Q = \frac{1}{\sigma}|J|^2 t, \qquad (1)$$

where σ is the electrical conductivity of the object and J is the eddy current density, t. It is known from Joule's Law that the heat generated increases as the eddy current density increases.

An infrared thermal imager can detect and record thermal radiation on the surface and display it in the form of a temperature value in the infrared image. Figure 2 shows a schematic of the operating principle of the infrared thermal imager.

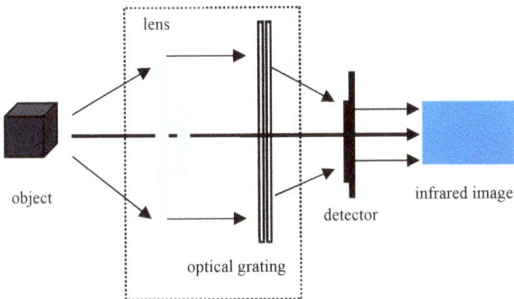

Figure 2. Operating principle of infrared thermal imager.

The different colors in the infrared image show the different temperatures on the surface of the object. A brighter region in the infrared image corresponds to a higher temperature on the surface of the detected object. The presence of an excitation source causes the infrared image collected in the experiment to differ from an image obtained under ideal conditions. However, most researchers have ignored the influence of the excitation source. An excitation source has a significant influence on an infrared image. The temperature of the surface in the presence of an excitation source is significantly higher than that of the surface far from the excitation source. We can find that there is a huge part whose size and shape is similar to the excitation source different from the background in the infrared image. Typically, the region influenced by the excitation source appears brighter in the infrared image, which means that the temperature is higher than the background without a defect underneath. If the defect is directly under the location of the excitation source on the surface, it is difficult to distinguish the defect. The area in the white box in Figure 3 is directly under the excitation source coil. There is a defect under the surface in the left box containing brighter colors. However, it is difficult to determine whether there is a defect inside the area in the right box, because its color is too similar to the background.

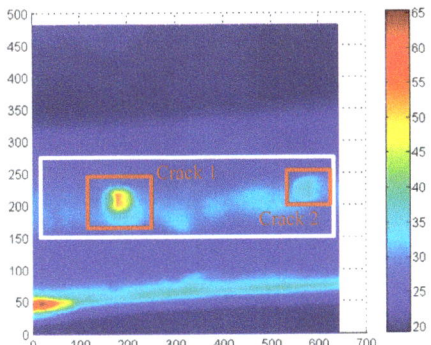

Figure 3. Infrared image.

The infrared image was meshed to analyze the temperature in the image. Figure 4 shows the temperature distribution of the image shown in Figure 3.

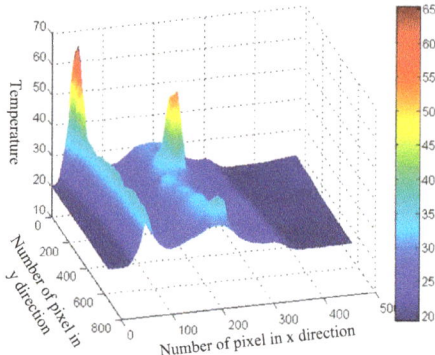

Figure 4. Meshed infrared image.

The temperature in the area directly beneath the excitation source was higher than that in other areas, causing a temperature swell to appear. The highest swell was caused by the defect in the left box in Figure 3. However, the swell caused by the defect in the right box in Figure 3 is not as clear because the excitation source is too large.

To eliminate the influence of the excitation source and make the defect profile clearer in the infrared image, several processing operations are proposed in this paper.

Several operations were applied to the pixels in the infrared image. We considered the value of a target pixel in the infrared image to be influenced by the neighboring pixels. The neighboring eight pixels were chosen to obtain the value of the target pixel.

2. Experiment Setup

In the infrared thermal imaging experiment, the specimen was heated to create different thermal distributions in the cracked and undamaged areas. A pulsed eddy current was applied to heat the material from the inside. In this way, the cracked area in the specimen produced considerably more heat than other areas. Generally, cracks distributed at different depths can be detected by NDT methods. The heat information at different depths depends on the rich frequency components during the pulsed eddy current excitation. Deep cracks could not be detected using infrared thermal imaging testing, because the heat transfer distribution became uniform over time as the heat was transferred from deeper regions to the surface. The experimental setup is shown in Figure 5.

Figure 5. Experimental setup.

Figure 5 shows that the excitation coil was restricted by the excitation controller. To detect deep cracks, the period of the square wave signal could be varied by the excitation controller. The specimen used in the experiment was a ferromagnetic material, and the cracks in it were artificial defects with depths of 0.1, 0.8, and 3 mm and width of 0.5 mm. The thermal imager and excitation controller could be simultaneously controlled by the computer (PC), where a 100 ms heating duration is selected for inspection. This heating time is long enough to elicit an observable heat pattern. The excitation coil is 60 mm width placed above the specimen, and the distance between the thermal imager and specimen is about 600 mm. The signal flow chart in the experiment is shown in Figure 6.

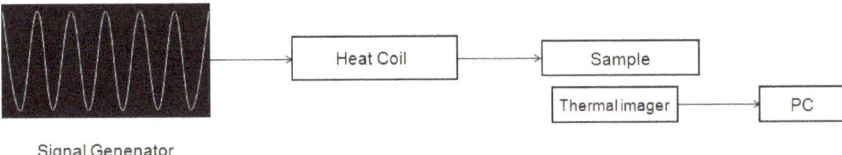

Figure 6. Signal flow chart.

3. Proposed Algorithm

The coordinate of the selected pixel is (i, j). The pixel value is denoted as $F(i, j)$. The values of the neighboring eight pixels with the specified step-sizes are as follows: $F(i-s, j-s)$, $F(i-s, j)$, $F(i-s, j+s)$, $F(i, j-s)$, $F(i, j+s)$, $F(i+s, j-s)$, $F(i+s, j)$, and $F(i+s, j+s)$, which can be expressed in matrix form as follows:

$$P = \begin{Bmatrix} F(i-s, j-s); \\ F(i-s, j); \\ F(i-s, j+s); \\ F(i, j-s); \\ F(i, j+s); \\ F(i+s, j-s); \\ F(i+s, j); \\ F(i+s, j+s); \end{Bmatrix}, \quad (2)$$

The model of the pixels in the infrared image is shown in Figure 7. The pixel value of a selected pixel, F(i,j) (yellow box in Figure 7), is considered to be related to the neighboring eight pixels, whose step-sizes are 2 (green boxes in Figure 7). For convenience, the neighboring pixels are expressed by a_i, arranged in clockwise order from the top left corner.

Figure 7. Model of pixels in infrared image.

To find the relationships between $F(i,j)$ and the neighboring pixels, we subtract the values of neighboring pixels from the value of $F(i,j)$ and arrange them in a column matrix, L, as follows:

$$L = \begin{Bmatrix} X(i,j) - a_1 \\ X(i,j) - a_2 \\ X(i,j) - a_3 \\ X(i,j) - a_4 \\ X(i,j) - a_5 \\ X(i,j) - a_6 \\ X(i,j) - a_7 \\ X(i,j) - a_8 \end{Bmatrix}, \tag{3}$$

The differences between $F(i,j)$ and the neighboring pixels shows the correlations between them. The correlation matrix, R, representing the correlation between the neighboring pixels is obtained from the difference matrix, L, as follows:

$$R = L \times L^T, \tag{4}$$

The differences between $F(i,j)$ and the neighboring pixels are denoted as Δ_1–Δ_8. After calculating the correlation between $F(i,j)$ and the neighboring pixels, the value of $F(i,j)$ is calculated as follows:

$$F(i,j) = \sqrt{\sum_{i=1}^{8}\sum_{j=1}^{8} \Delta_i \Delta_j a_i a_j}, \tag{5}$$

Equation (5) can be written as follows:

$$F(i,j) = \sqrt{P^T \times R \times P}, \tag{6}$$

The value of $F(i,j)$ obtained from the process above is not the exact value, as the sum of the correlation coefficients is not equal to 1. Thus, to obtain the real value of $F(i,j)$, the value above must be normalized as follows:

$$F(i,j) = \frac{F(i,j)}{(\sum_{i=1}^{n} \Delta_i)^2}, \tag{7}$$

The infrared image can finally be obtained by calculating the value of every pixel using the method presented above.

4. Results and Discussion

The infrared image analyzed is shown in Figure 3. The image was collected with the excitation source directly above the surface and two defects under it. The defect on the right side of the image is not clear.

By verifying the value of the step-size, we determined that the method was most efficient when the step-size was 8. Figure 8 shows the result when the step-size was 8. Compared with the initial infrared image, it is evident that the swell caused by the excitation source was eliminated.

Figure 9 shows the contrast between the processed infrared image and the initial image. The infrared image after processing was significantly clearer than the initial image. The bright area influenced by the excitation source was eliminated in the processed infrared image. Furthermore, the bright area in the right box was caused by the defect inside the object became clearer.

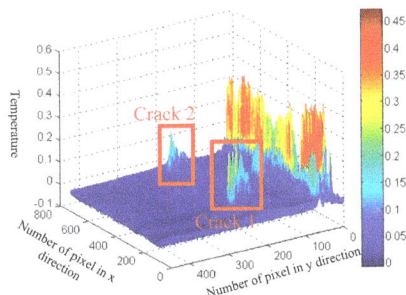

Figure 8. Meshed infrared image after processing.

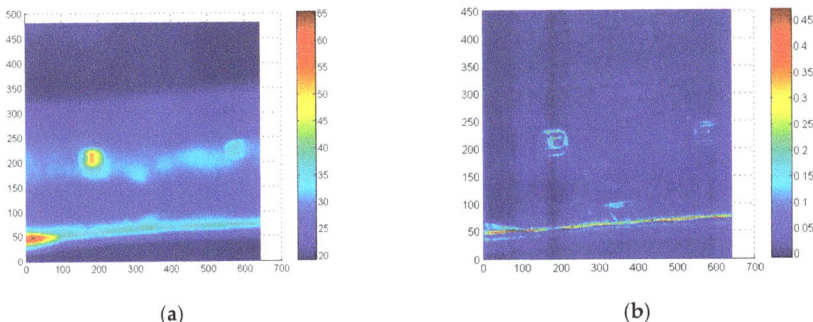

Figure 9. Comparison of initial and processed images: (**a**) Initial infrared image; (**b**) Infrared image after processing.

The swell in the infrared image caused by the excitation source was eliminated by using the algorithm provided in this paper. Furthermore, the outlines of the defects were made more distinct. The method presented in this paper improved the defect detection accuracy by enhancing the two crack images, and most importantly, it reduced the background noise which could increase the credibility of crack estimation compared with the original detected infrared image.

It should be noted that this algorithm is designed to reduce the background noise imported through the temperature of the environment and the excitation. Furthermore, in the pulsed eddy current infrared imaging test, multiple cracks will influence the distribution of the temperature so that the low temperature area of a crack will be covered in the background. This is because the thermal imager should adapt to the temperature in the whole vision area to make sure the image is clear. Thus, the proposed method is motivated by this problem. From the results, it can be inferred that when multiple cracks exist in the visible area, this method could reduce the influences produced by the thermal transforming. At stable temperatures, because the temperature between every two crack is similar and smooth but the temperature in the crack area is higher than that of the other areas, this method could be used.

5. Conclusions

A novel algorithm was proposed to enhance the infrared images by performing mathematical operations using pixel values with a specific step-size for the detection of welding defects. The pixel value was calculated using the neighboring pixels' values and the correlation between them. The relationship between the neighboring pixels was inferred from the differences between the target and neighboring pixel values. The correlation coefficients between the neighboring pixels were arranged in a matrix. The elements were multiplied by the corresponding pixel values and were subsequently summed and normalized, yielding a final pixel value. The method proposed in this paper represents each pixel value using the neighboring pixels, eliminating swells in an infrared image due to an excitation source, which make the detection of defects significantly more difficult. The outline of the defect was found to be clearer when the processing was performed using the above-mentioned method.

Author Contributions: H.L. contributes to most of the results including modeling construction and analysis of the results, and he also completes the English writing of this paper. Y.Y. proposed the main idea of this article. L.L. and B.L. offer some help in processing data and draw diagrams.

Funding: This work is financial supported by the Fundamental Research Funds for the Central Universities under Grant ZYGX2018J067 and the Nature Science Foundation of Guangdong Province under Grant 2018A030313893.

Conflicts of Interest: The authors declare no conflicts of interest.

References

1. Ebrahimkhanlou, A.; Athanasiou, A.; Hrynyk, T.D.; Bayrak, O.; Salamone, S. Fractal and Multifractal Analysis of Crack Patterns in Prestressed Concrete Girders. *J. Bridge Eng.* **2019**, *24*, 04019059. [CrossRef]
2. Ebrahimkhanlou, A.; Salamone, S.; Ebrahimkhanlou, A.; Azad, A.R.G.; Kreitman, K.; Helwig, T.; Williamson, E.; Engelhardt, M. Acoustic emission monitoring of strengthened steel bridges: Inferring the mechanical behavior of post-Installed shear connectors. In Proceedings of the Nondestructive Characterization and Monitoring of Advanced Materials, Aerospace, Civil Infrastructure, and Transportation XIII, Denver, CO, USA, 4–7 March 2019.
3. Ebrahimkhanlou, A.; Choi, J.; Hrynyk, T.D.; Salamone, S.; Bayrak, O. Detection of the onset of delamination in a post-Tensioned curved concrete structure using hidden Markov modeling of acoustic emissions. In Proceedings of the Sensors and Smart Structures Technologies for Civil, Mechanical, and Aerospace Systems 2018, Anaheim, CA, USA, 26–30 April 2018.
4. Tian, L.; Cheng, Y.; Yin, C.; Ding, D.; Song, Y.; Bai, L. Design of the MOI method based on the artificial neural network for crack detection. *Neurocomputing* **2017**, *226*, 80–89. [CrossRef]
5. Park, H.; Choi, M.; Park, J.; Kim, W. A study on detection of micro-Cracks in the dissimilar metal weld through ultrasound infrared thermography. *Infrared Phys. Technol.* **2014**, *62*, 124–131. [CrossRef]
6. Rodríguez-Martin, M.; Lagüela, S.; González-Aguilera, D.; Arias, P. Cooling analysis of welded materials for crack detection using infrared thermography. *Infrared Phys. Technol.* **2014**, *67*, 547–554. [CrossRef]
7. Dapeng, C.; Hongxia, M.; Zhihe, X. Infrared Thermography NDT and Its Development. *Comput. Meas. Control* **2016**, *4*, 1–6.
8. Plotnikov, Y.A.; Winfree, W.P. Advanced Image Processing for Defect Visualization in Infrared Thermography. *Proc. SPIE* **1998**, *3361*, 331–338.
9. Liu, Z.P.; Hu, L.H.; Zhou, J.M.; Cai, L. Evaluation of Surface Defect Area in Metal Based on Infrared Thermal Image. *Appl. Mech. Mater.* **2014**, *530*, 171–174. [CrossRef]
10. Min, Q.X.; Feng, F.Z.; Wang, P.F.; Zhang, C.S.; Zhu, J.Z. Recognition of contact interface defect in metal plate based on pulsed phase thermography. In Proceedings of the Prognostics & System Health Management Conference, Beijing, China, 21–23 October 2015.
11. Fan, C.F.C.; Sun, F.S.F.; Yang, L.Y.L. A general quantitative identification algorithm of subsurface defect for infrared thermography. In Proceedings of the 2005 Joint 30th International Conference on Infrared and Millimeter Waves and 13th International Conference on Terahertz Electronics, Williamsburg, VA, USA, 19–23 September 2005.
12. Meola, C.; Carlomagno, G.M. Recent advances in the use of infrared thermography. *Meas. Sci. Technol.* **2004**, *15*, 27. [CrossRef]
13. Maldague, X. Applications of infrared thermography in nondestructive evaluation. In *Trends in Optical Nondestructive Testing*; Elsevier Science: Amsterdam, NL, USA, 2000; pp. 591–609.
14. Avdelidis, N.P.; Hawtin, B.C.; Almond, D.P. Transient thermography in the assessment of defects of aircraft composites. *Ndt E Int.* **2003**, *36*, 433–439. [CrossRef]
15. Wang, X. Transient Thermography for Detection of Micro-Defects in Multilayer Thin Films. Ph.D. Thesis, Loughborough University, Loughborough, UK, 2017.

16. Chen, D.; Zhang, X.; Zhang, G.; Zhang, Y.; Li, X. Infrared Thermography and Its Applications in Aircraft Non-destructive Testing. In Proceedings of the 2016 International Conference on Identification, Information and Knowledge in the Internet of Things (IIKI), Beijing, China, 20–21 October 2016.
17. Li, H.J.; Lin, J.G.; Mei, X.; Zhao, Z. Infrared Image Denoising Algorithm Based on Adaptive Threshold NSCT. In Proceedings of the 2008 Congress on Image and Signal Processing, Sanya, China, 27–30 May 2008.
18. Ibarra-Castanedo, C.; Gonzalez, D.; Klein, M.; Pilla, M.; Vallerand, S.; Maldague, X. Infrared image processing and data analysis. *Infrared Phys. Technol.* **2004**, *46*, 75–83. [CrossRef]
19. Wang, J.; Hong, J. A new self-Adaptive weighted filter for removing noise in infrared images. In Proceedings of the 2009 International Conference on Information Engineering and Computer Science, Wuhan, China, 9–20 December 2009.
20. Voronin, V.; Tokareva, S.; Semenishchev, E.; Agaian, S. Thermal Image Enhancement Algorithm Using Local and Global Logarithmic Transform Histogram Matching with Spatial Equalization. In Proceedings of the 2018 IEEE Southwest Symposium on Image Analysis and Interpretation (SSIAI), Las Vegas, NV, USA, 4–10 April 2018.
21. Bai, J.; Chen, Q.; Wang, X.; Qian, W. Contrast enhancement algorithm of infrared image based on noise filtering model. *Infrared Laser Eng.* **2010**, *39*, 777–780.
22. Liu, T.; Zhang, W.; Yan, S. A novel image enhancement algorithm based on stationary wavelet transform for infrared thermography to the de-bonding defect in solid rocket motors. *Mech. Syst. Signal Process.* **2015**, *62*, 366–380. [CrossRef]
23. Yin, A.; Gao, B.; Yun Tian, G.; Woo, W.L.; Li, K. Physical interpretation and separation of eddy current pulsed thermography. *J. Appl. Phys.* **2013**, *113*, 064101. [CrossRef]
24. He, Y.; Pan, M.; Luo, F. Defect characterisation based on heat diffusion using induction thermography testing. *Rev. Sci. Instrum.* **2012**, *83*, 104702. [CrossRef]

© 2019 by the authors. Licensee MDPI, Basel, Switzerland. This article is an open access article distributed under the terms and conditions of the Creative Commons Attribution (CC BY) license (http://creativecommons.org/licenses/by/4.0/).

MDPI
St. Alban-Anlage 66
4052 Basel
Switzerland
Tel. +41 61 683 77 34
Fax +41 61 302 89 18
www.mdpi.com

Applied Sciences Editorial Office
E-mail: applsci@mdpi.com
www.mdpi.com/journal/applsci

www.ingramcontent.com/pod-product-compliance
Lightning Source LLC
LaVergne TN
LVHW070641100526
838202LV00013B/850